Adventures in Earth Science Student Practical Manual

Dr. Peter T. Scott

First released 2018 all rights reserved Felix Publishing
Felix Publishing 2018

www.felixpublishing.com.au
email: info@felixpublishing.com
Print copies available from publisher.

ADVENTURES in EARTH SCIENCE - STUDENT PRACTICAL MANUAL

The textbook is also available as a series of smaller books:

- Exploration Science (Field Geology and Mapping)
- Riches from the Earth (Minerals, Mining & Energy)
- Changing the Surface (Erosion and Landscapes)
- Rocks - Building the Earth
- Fossils - Life in the Rocks
- A Dangerous Planet (Earth Hazards)
- Through Sea and Sky (Oceanography and Meteorology)
- Beyond Planet Earth (Astronomy)

2018 digital book release
ISBN: 978-1-925662-15-3
Print Edition
ISBN: 978-1-925662-14-6
Author: Dr. Peter T. Scott

All illustrations, photographs and videos by the author unless stated
Cover photo: Andrew Scott of AJS Creative

Registration:
Thorpe-Bowker +61 3 8517 8342
email: bowkerlink@thorpe.com.au

No part of this publication may be reproduced, stored in a retrieval system, or transmitted in any form or by any means, electronic, mechanical, photocopying, recording or otherwise, without the prior written permission of the publisher.

© All rights reserved Felix Publishing 2018

About the Author

Dr. Peter T. Scott, Paradise Bay, Antarctica 2011

Dr. Peter Scott is an award-winning Earth Science teacher of over forty years' experience Secondary schools. He holds a bachelor's degree, two Master's degrees and a Doctorate in both the fields of Science and Education and has been an advisor in Science Education to several State governments. Apart from his own research exploration, he has also travelled extensively and has visited many places of interest including North and South America, Antarctica, North Africa, the volcanic islands of the Pacific, Asia, northern Europe, and Australia.

Table of Contents

Introduction   1.
Writing Practical Reports; Safety in the Classroom Laboratory; Safety in the Field.

Chapter 1: Exploration Science   8.
1.1 Observation of pyrite crystals
1.2 Measuring feldspar crystals
1.3 Using a Topographical Map.

Chapter 2: Rock-forming Minerals   16.
2.1 Some Common Rock-forming Minerals
2.2 Growing Crystals
2.3 Measuring Specific Gravity
2.4 Geochemistry

Chapter 3: Igneous Rocks - The Beginning   26.
3.1 Some Common Igneous Rocks

Chapter 4: Sedimentary Rocks   28.
4.1 Some Common Sedimentary Rocks
4.2 Modal Analysis of Sediment
4.3 Porosity and Permeability

Chapter 5: Metamorphic Rocks   36.
5.1 Some Common Metamorphic Rocks

Chapter 6: Weathering and Erosion   38.
6.1 Weathering
6.2 Introduction to Stream Table Experiments
6.3 Simple Soil Testing Experiments

Chapter 7: Landforms   46.
7.1 Karst Simulation
7.2 The Shape of Rivers
7.3 Stereopairs and Landforms

Chapter 8: Fossils - Life in the Rocks   52.
8.1 Some Common Fossils
8.2 Correlation Exercise

Chapter 9: Economic Minerals and Mining   56.
9.1 Some Common Economic Minerals

Chapter 10: Processing the Mined Ore   59.
10.1 Making Copper from a Copper Ore

Chapter 11: Fuels and Energy   61.
11.1 Fossil Fuels
11.2 Energy Content of Fuels

Chapter 12: Exploring the Seas   66.
12.1 Introduction to Navigation
12.2 Mapping the Depths

Chapter 13: The Dynamic Earth 73.
13.1 Hooke's Law
13.2 Deformation and Temperature
13.3 Compressional Structures

Chapter 14: Earthquakes 79.
14.1 Introduction to Seismology
14.2 Locations of Some Major Earthquakes

Chapter 15: Volcanoes 83.
15.1 Shape of a Volcano
15.2 Locations of Some Major Volcanoes

Chapter 16: Geological Maps 87.
16.1 Mapping Inclined Beds
16.2 Mapping Faulted Beds
16.3 Mapping Folded Beds
16.4 Mapping Igneous Intrusions
16.5 Mapping Unconformities
16.6 Geological History

Chapter 17: Moving Sea and Wind 106.
17.1 Relative Humidity
17.2 The Aneroid Barometer
17.3 Water Density
17.4 Convection Currents

Chapter 18: Planet Earth 115.
18.1 The Rotation of the Earth
18.2 Size of the Earth

Chapter 19: The Moon 120.
19.1 The pathway and Shape of the Moon
19.2 The Size of the Moon
19.3 The Main Features of the Moon

Chapter 20: A Matter of Perspective 127.
20.1 Planetary Motion
20.2 Elliptical Orbits

Chapter 21: More on Telescopes 133.
21.1 The Refracting Telescope
21.2 The Reflecting telescope
21.3 Spectroscopy

Chapter 22: More on the Solar System 144.
22.1 The Solar Spectrum
22.2 Observing Sunspots
22.3 Introduction to the Planets
22.4 The Orbit of Mars

Chapter 23: Beyond to the Stars 152.
23.1 Constellations – Southern Hemisphere
23.2 Constellations – Northern hemisphere
23.3 Finding Distance by Parallax

Introduction

1. Why "Adventures" in Earth Science?

The study of the Earth IS an adventure! Studying Earth Science involves:

- **Exploration** – whether it be in remote areas or near places of habitation, field work usually involves the breaking of new ground (forgive the pun!). Often in the more remote locations, field work might be over ground which has never been as thoroughly explored as in your study.

- **Exciting Places** – usually outside of the person's usual habitation range and often in very remote places in different parts of the world, in different physical environments such as deep underground (mines and caves), on high mountain ranges and in different climatic conditions (open oceans, hot and cold deserts and jungles).

Photo: Paradise Bay, Antarctica

- **Meeting Interesting People** – who are often happy to share their own experiences and culture, not to mention their home, food and local knowledge. One meets a surprising number of people, mostly good people, such as fellow Earth Scientists and students, field workers in the Earth Industries, Government officials; local land-owners and indigenous peoples.

- **Exciting Research** – especially when working on a new project or studying a new location, Earth Science research, as with all scientific studies can be especially motivating (sometimes to the point of happy obsession!). Just like a good detective story, there are things to find out, evidence to gather, step-by-step deductions to be made and finally a conclusion which answers the research question. Looking for gold or gems would be a simple example but it is when the study has a great number of research questions that the field and its companion laboratory work becomes exciting.

- **Studying at many levels** – whether it is a simple prospecting trip or a detailed study of comet movement in the sky over a long time period (NASA leaves this to gifted amateurs!), Earth Science activities can occur at any age and at every level from the school student amateur to the professional Scientist.

Whatever the nature of the study or the level of the student, the Earth is a dynamic and active place with exciting things to see. If Planet Earth is not adventurous enough, there is always the rest of the Universe.

2. Studying Earth Science

Earth Science covers a wide range of interesting topics, from volcanoes to cutting gems. The amount of mathematics, often common to Science subjects, has been reduced within the textbook; providing adequate analysis of large amounts of data, but there is still much content to cover in order that a true understanding of the Earth and beyond is obtained.

Because of the amount of content needed to be learned, it is highly recommended that students adopt sound attitudes and practices. These include:

- **Making good, concise summaries of each topic**. Whilst electronic devices such as computers and tablets are excellent for note taking, there is a tendency in modern education for students (and teachers) to upload or cut/paste previously prepared notes. These are good for gathering information as a primary source, but eventually a certain (minimal) form of summary will be required for study purposes and committing a basic set of notes to memory. After all, one must have some information stored in the brain (as well as on Hard Drive) to be able to use when using what has been learned. The best way is the ancient method of **reading, analysing and extracting the main ideas** and then **writing them down on paper** as a study summary. Doing this on a computer screen has some usefulness but it is not as good for learning as the hand-eye coordination which occurs with physical writing on note paper as a study sheet. Students should organize these sheets so that there is one page per major topic. The use of simple diagrams, charts and lists is an advantage to learning. For long sequences which should be known at all times and not simply retrieved from a data bank on demand, mnemonics are most useful. These may take a simple form of using the first letter of each word in the sequence to make another, simpler word which forms part of a crazy sentence. The crazier, the better. This is then learnt off by rote (try writing it out 100 times!) with the appropriate mental connections made to the real words and sequence. For example, consider the main planets in order from the Sun – Mercury, Venus, Earth, Mars, Jupiter, Saturn, Uranus, Neptune. This could be remembered by "My Very Elegant Mother Just Sat Upon Nettles".

- **Good study habits** and time management are not natural processes for most students, even those of mature age. A time-management grid or fixed calendar of regular study and assignment preparation could be suggested by the teacher. This would also include the necessary breaks for leisure time and social activities. Such a timetable should be flexible and include such well-known concepts that there should be social/leisure time immediately after classes; study is best done just before sleeping; that a variety of different topics are best done in small amounts rather than a massive amount on the same topic; and that short (written) summaries of the nightly study are best done last. The latter concept does waste a lot of paper but the hand-eye coordination is most useful for memory retention.

- **Completing practical work and assignments on time.** Never leave due work until the last minute. Students should be taught to plan a sequence of preparation e.g. gather notes by library/Internet research and reading, summarizing these notes in a cohesive form, planning the structure of the main work, then writing the assignment in one or several stages so that it is completed well before the due date. Assignments and practical work which occur regularly and often should be started as soon as possible after they have been set. Interruptions due top personal life will occur and if these interfere with the submission of work, extra time should be given by the teacher if they are valid.

- **Noting down ideas**, words etc. not understood should become a regular habit, especially as data recording whilst working is part of the Earth Science method. Difficult questions, new observations and unknown concepts are the source of further research by asking questions. Teachers should encourage their students to always ask questions about what they see and do. There should be a classroom understanding that there is no such thing as either an awkward or silly question. We learn by questioning – even established ideas.

- **Students should work independently and be accountable for their own work.** Although they may have to work and gather data in small groups, students must understand that the final outcome of this work should be their own. It is very easy using electronic media to copy and manipulate other people's original work. Even without computer programs to analyse student work for such plagiarism, it is usually obvious to the trained teacher to see that the level of understanding, terms used and data outlines is not typical of the age nor academic level of the student. Originality, no matter how simple, is the key to a good assignment.

3. Writing Practical Reports

Each school Faculty will have its own view on the recording of any formal practical work completed by the student. It would be hoped that any formal writing, submission and assessment of student practical report will follow a sequence common to that employed later in their studies as post-graduate students or professional Earth Scientists. This sequence is part of the recording phase of the Scientific Method and should be done so that the practical work or experiment can be read and vetted by others and used as a stimulus for further new experimentation and study. A good report or thesis on any research project should follow a logical sequence generate more questions than it answers.

Scientists, like other managers must write down records of their work and present reports. Earth Scientists working for private companies or government agencies write regular reports to their Directors. Research Scientists publish papers (detailed articles for well-known scientific magazines), present seminars and write books about their work. Students write practical reports or theses for schools and universities. Regardless of the type of report, they all follow a similar system: there is an AIM or intent; a detailed account of the METHOD used; some outline of the findings of the research, or their RESULTS; and eventually a CONCLUSION to state what the research proved, how accurate it was and what good it will do.

An example of a sequence of the main components of any written practical reports may be:

1. Giving a **HEADING** and **DATE** (and if required by the teacher, names of co-workers)

2. The **AIM** is written in FUTURE tense e.g. "To ...." There may be one or two aims but never too many. The Aim shows others what you are setting out to show.

3. There may be a need to put in a brief list of **MATERIALS** to show what was used;

4. The **PROCEDURE** is always written in THIRD PERSON (i.e. impersonal e.g. First Person is "I", Second Person is "We" and Third Person does not use personal pronouns at all.) and in PAST TENSE as though the student has already completed the work (because when it is read it **will** have been completed). The Method is best written in point form (some prefer a Flow Diagram but this can be difficult to construct if the method is complex) e.g.

> "1. The specimen was carefully examined with a hand lens;
> 2. Noting its overall shape, an outline was drawn;
> 3. Internal features in the shape were again examined;
> 4. These features were then drawn within the outline;
> 5. Colour and shading was applied; and
> 6. The original specimen's size was compared to that of the sketch to derive a scale in size (e.g. x5)"

5. **OBSERVATIONS & DATA** will include (in third person, past tense) descriptions of ALL observations found using all senses and/or instruments used in detection or any measurement; any data collected; calculations made (some teachers prefer a separate CALCULATIONS section after results); and whenever possible, a DRAWING or SKETCH of any specimen, apparatus or major feature. In Earth Science, drawings of laboratory apparatus are done as in the other sciences such as Chemistry, as 2-dimensional pencil (black ink only if students are good at drawing) using rulers and printed labels in normal ink. Other sketches e.g. of specimen should be done in 3-dimensions in pencil and later coloured pencils and a SCALE of the size (e.g. x1 or x 2 etc.) given. For electronic submissions of work, students are encouraged to use appropriate software for writing the report, doing the artwork and inserting their **own** photographs. If hand-sketched drawings made during the classroom activity are to be included, these should be photographed or scanned **after** completion taking care with colour, size and contrast of the image. Students' own videos may also be included to illustrate important activity but students who have poor computing skills still should be encouraged to submit written reports. The teacher should devise an assessment system in grading reports so that written or electronic submissions are given fair comparisons.

6. **QUESTIONS** if there are specific questions to be answered, it is advisable to also re-write the question and then give its answer.

7. The **CONCLUSION** gives the answer to the aim, any errors encountered and suggestions for improvements. Sometimes, more substantial reports may also have a final section which refers to possible new research.

The first activity (Observation of Pyrite Crystals) is given to you as an example in the way reports will be written in the future. Future activities will have instructions which must be rewritten in the standard manner.

8. Safety in the Classroom Laboratory

Experience has shown that some Secondary and Tertiary students are uncoordinated and socially irresponsible. For this reason, they and rest of the group need to be protected by the application of a set of firm but fair Safety Rules. These should be simple, able to be comprehended by the student and practiced. It is also a good idea to have them posted within the classroom. In time, most groups get into a working pattern and so The Rules become simply standard learning procedures and so a good, working social atmosphere develops. An example of such rules is given on the next page:

SAFETY in the LABORATORY

The following policy is in place to ensure that the laboratories are safe places for students to work in. Respect for self, others and property is first priority. Please ensure that:

1. Safety apparel is to be worn at all times – aprons, good enclosed shoes, eye glasses etc.;
2. you do not enter a laboratory without teacher supervision;
3. unauthorized experiments are strictly forbidden. Variations in procedure must be approved;
4. food and drink are not consumed in the laboratory;
5. movement and noise should be kept to a minimum as distractions cause accidents;
6. items and liquids are not thrown at any time;
7. you do not touch, taste or smell chemicals and minerals only as directed by the teacher;
8. spillages, breakages and accidents are reported immediately;
9. you know where all safety switches and equipment are located and how they are used;
10. the work area is clean and tidy. When finished, clean apparatus and return with chemicals etc. to appropriate place. Wipe down laboratory bench and wash hands.

In addition, use of electronic equipment such as computers, tablets and cell phones should be appropriate for the learning environment and secondary to the teaching process. Some schools ban the use of cell phones but with a few simple courtesies they can be useful in the classroom.

9. Safety in the Field

This requires more vigilance than in the protected laboratory environment. In the field, away from the classroom, there are other external influences such as:

- the climate, which could be excessive in its heat or cold, windy, wet or subject to sudden change;

- the terrain, which may be steep, rugged, slippery, loose, and full of ravines, caves or mine shafts;

- the vegetation, which may be dense, tangled, thorny and sometimes poisonous; and

- the wildlife, may be dangerous if disturbed or generally a nuisance (such as flies or mosquitos).

Students should also take precautions such as:

- not to wander about and but keep with the group. There is real danger of becoming lost, falling into ravine or mine shaft or other dangers;

- being prepared for the trip, especially about such necessities as:

    CLOTHING - especially adequate footwear, head covering, suitable clothing for the climate and sturdy boots;
    WATER and FOOD as appropriate for the trip;
    SPECIAL SAFETY GEAR (e.g. walking poles);
    INSECT REPELLANT and SUN PROTECTION;
    STUDY ITEMS such as Excursion Guide, notepad and pencil.

- forewarn the teacher about any special needs such as allergies and other problems involving outdoor activities; and

- do not indulge in foolish behaviour such as throwing stones or being a distraction.

Dangerous obstacles should be avoided and only safe, secure tracks should be used. Good navigation is essential and the proposed route and time of return should be left with any local authority such as local police or park rangers as well as with the school authorities. A typical set of Field Safety Rules are:

FIELD RULES

When in the field:

1. Listen to all instructions - especially about specific local hazards;

2. Keep with the group - do not wander;

3. Do not enter bodies of water unless told to do so;

4. Do not enter old mines or industrial workings unless told to do so, then with caution;

5. Do not climb cliffs nor stand under or near unstable rocks;

6. Do not throw any objects, especially hammers nor rocks;

7. Watch your step, especially on slopes and in close vegetation;

8. Wear appropriate field clothing at all times. Be prepared for sudden changes in the weather (rain, cold/heat). Carry a waterproof jacket;

9. Watch out for traffic when on or near roads and railway cuttings;

10. Carry own water and some food;

11. Keep movement and noise to a minimum;

12. Do not use cell phones inappropriately. Headphones are not allowed. Take own care of cameras.

Chapter 1: Exploration Sciences

EXPERIMENT 1.1   **OBSERVATION OF PYRITE CRYSTALS**   Time: One Lesson

AIM: To use a hand lens or binocular microscope to observe, sketch and describe pyrite crystals

MATERIALS: Pyrite crystal block   hand lenses or binocular microscopes   coloured pencils

BACKGROUND:

Pyrite is also called Fool's Gold because in fine veins it looks like real gold. CAREFUL OBSERVATION is needed to distinguish it from the real thing. Sketching the specimen to scale (usually about x3 or x4) is a good way to look for the special features of this mineral.

PROCEDURE: (NOTE: THIRD PERSON, PAST TENSE, and POINT FORM)

PART A: SETTING UP THE MICROSCOPE

1. The microscope was carefully removed from its box and the box put aside;
2. The large black knob on the upright rod was loosened and the whole body of the microscope was moved up to near the top of the rod;
3. The eyepieces were adjusted so that one complete circle was seen;
4. The specimen was placed in the centre of the stage and the eyepiece housing was moved down to near the specimen using the black focusing knobs;
5. Looking down the eyepieces the focusing knob was adjusted UPWARD until a clear view was seen.

PART B: OBSERVING DETAIL

6. The overall shape of the specimen was carefully examined and compared with common shapes (e.g. cubes, rectangles, triangular prisms etc.);
7. Details of COLOUR, SHADING and INTERNAL SHAPES (i.e. shapes within the specimen and how they are placed together) were observed as were any INTERNAL DETAILS such as lines, bubbles, spots, cracks etc....any regular patterns;

PART C: SKETCHING THE SPECIMEN

8. A circle was drawn (about one third of a page) to represent the FIELD OF VIEW of the microscope and a rule was placed under the microscope to measure the actual field of view (e.g. about 10 mm). This was written near the circle to give SCALE;
9. Thoughts about what was seen (shape, colour, position of parts etc.) were revised and sketched using very light pencil;
10. Detail was re-observed and included in the sketch and the outline was re-done with heavier pencil (or black ink).

EXPERIMENT 1.1 continued

DATA and OBSERVATIONS:

Copy a circle into which the sketch can be made (DO NOT DRAW IN THIS BOOK):

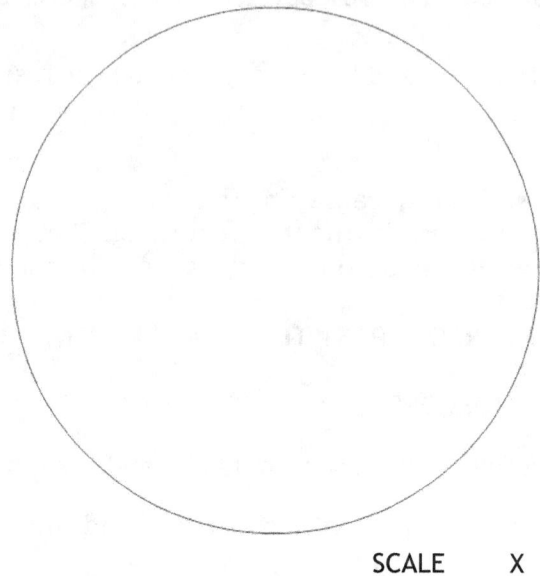

SCALE   X

Do a detailed sketch in colour and add labels for distinctive features. Also give the scale.

Describe the main observations:

CONCLUSIONS

1. What are the main distinguishing features of Pyrite?
2. Comment on factors which might limit the accuracy of observation.
3. Why is it important to have an accurate scale in drawing?
4. Other general conclusions (if required)

(NOTE: Unlike the first Activity 11.1, students are required to rewrite their reports from this document as AIM, METHOD, RESULTS, and CONCLUSION using THIRD PERSON, PAST TENSE, and POINT FORM. MATERIALS and BACKGROUND may not be required in your report).

EXPERIMENT 1.2   Time: One Lesson

MEASUREMENT of FELDSPAR CRYSTALS

Use a photomicrograph (photo taken through a Geological Microscope) to estimate the size of feldspar crystals) and determine any relationship between size of crystals and depth of their formation.

MATERIALS: Three photomicrographs (each x 25 magnification) of separate specimens of BASALT rock, rulers, and calculators as required.

BACKGROUND:

Three photomicrographs (each x 25 magnification) taken of thin-sections of separate specimen of a BASALT rock found at different depth. Each photomicrograph has been made by cutting a piece of the basalt to a thin slice which is then glued to a glass slide and ground down until transparent. It is then viewed in polarized light and photographed

PROCEDURE:

For EACH photomicrograph measure a good number (say 20) of the LENGTHS of the feldspar crystals (seen as long shapes) in EACH of the three photographs. The scale of each photo has been made up to the real size of the crystals (i.e. what you measure with your ruler is the correct size)

Plot your data in a Table for each of the photographs relating sizes and depths:

e.g.	Photo 1 (Depth =)	Photo 2 (Depth =)	Photo 3 (Depth =)
	measurement 1 mm measurement 2 etc. etc. (20 measurements) Average = mm	measurement 1 mm measurement 2 etc. etc. Average = mm	measurement 1 mm measurement 2 etc. etc. Average = mm

Use your knowledge of mathematics to find an AVERAGE size of the crystals for each photomicrograph and estimate the ERROR of MEASUREMENT for your final value.

Make a sketch of one of the photomicrographs as a representation of those given. Provide a scale for your sketch.

EXPERIMENT 1.2 continued

Photomicrograph 1. (Depth 2 metres below surface)   Photomicrograph 2. (10 metres below surface)

Photomicrograph 3. (20 metres below surface)

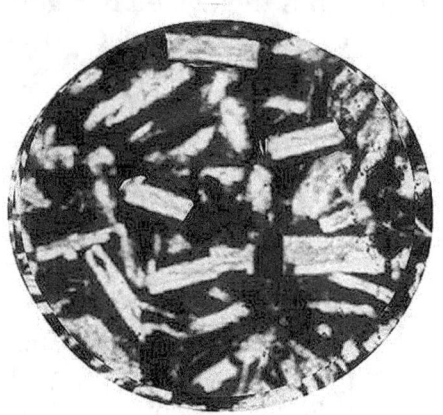

CONCLUSIONS:

1. What was the average size of the crystals for each of the depths and their error of measurement?

2. What TYPE of error of measurement is this?

3. Is there a relationship between crystal size and depth of their formation? If so, what is it? Can you express this relationship as a mathematical expression?

4. Are your measurements made on the photographs a true representation of the sizes of the crystals? Why? How can you improve on the method of **randomly** sampling which crystals are to be measured?

5. What are some other conclusions about the size of crystals or method used?

6. How could you further test your hypothesis about size of crystals and depth?

RESEARCH (Optional at the teacher's discretion):
What are feldspar crystals? How do they form in rocks such as basalts? Where would VERY large crystals of feldspar (say a metre long) form? What use is feldspar in daily life?

EXPERIMENT 1.3

One or two Lessons

USING A TOPOGRAPHICAL MAP

AIM: To use a topographical map to locate and refer to objects or features, measure distance and draw a simple cross-section to scale.

MATERIALS: Pencils, rulers, grid/graph paper, paper, string

BACKGROUND:

Topographical maps are still most useful despite the dependence on GPS devices. They can be used in obtaining a general idea of a research area and some of its features. Electronic devices can sometimes fail in some remote areas. Aerial photos showing the real (as opposed to mapped) features are also very useful.

PROCEDURE:

(a) General Features of a Topographic Map.

Look at the general features of the topographical map on the next page:

QUESTIONS:

1. What is the scale of this map? What does this mean in reality?
2. What are the GEOGRAPHICAL COORDINATES (i.e. Latitude & Longitude of this area?
3. Use an atlas (printed or electronic) to find where this place is located.
4. What is the MAGNETIC VARIATION of Utopia Point? (as degrees East)
5. List the GRID NUMBERS which represent:
    a. the EASTINGS and
    b. the NORTHINGS.
6. Use the SCALE and a ruler to measure the length of the JETTY at UTOPIA POINT.
7. Use a piece of string to estimate the length of the coastline shown on the Map.
8. Within which GRID SQUARE is HOSTPITAL POINT?
9. What objects are located at GRID REFERNCE:
    a. 393583
    b. 428590
10. What is the GRID REFERNCE for:
    a. the end of the JETTY at Utopia Point
    b. the lone hill (118 m high) in the far northeastern part of the map
11. What is the CONTOUR INTEVAL of this map?
12. What is the feature between grids 413576 and 414564?

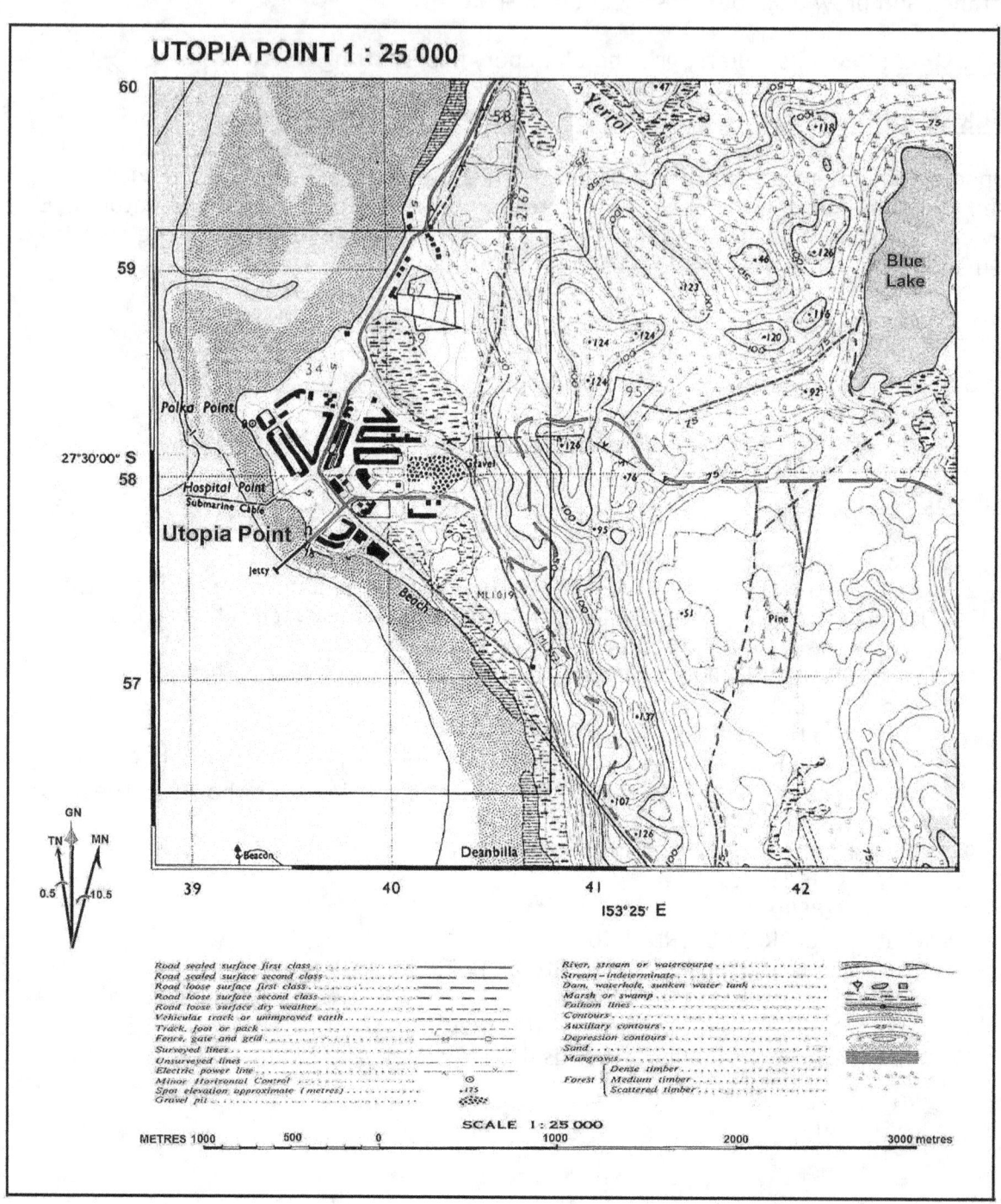

EXPERIMENT 1.3 continued

PROCEDURE (continued):

(a) Drawing a cross-section.

Whilst there are computer programs which will draw map cross-sections, it is still useful to know how it is done and to understand the concepts of VERTICAL EXAGGERATION, GRADIENTS and the basic shapes of LAND FEATURES.

Consider the following large scale part of a map:

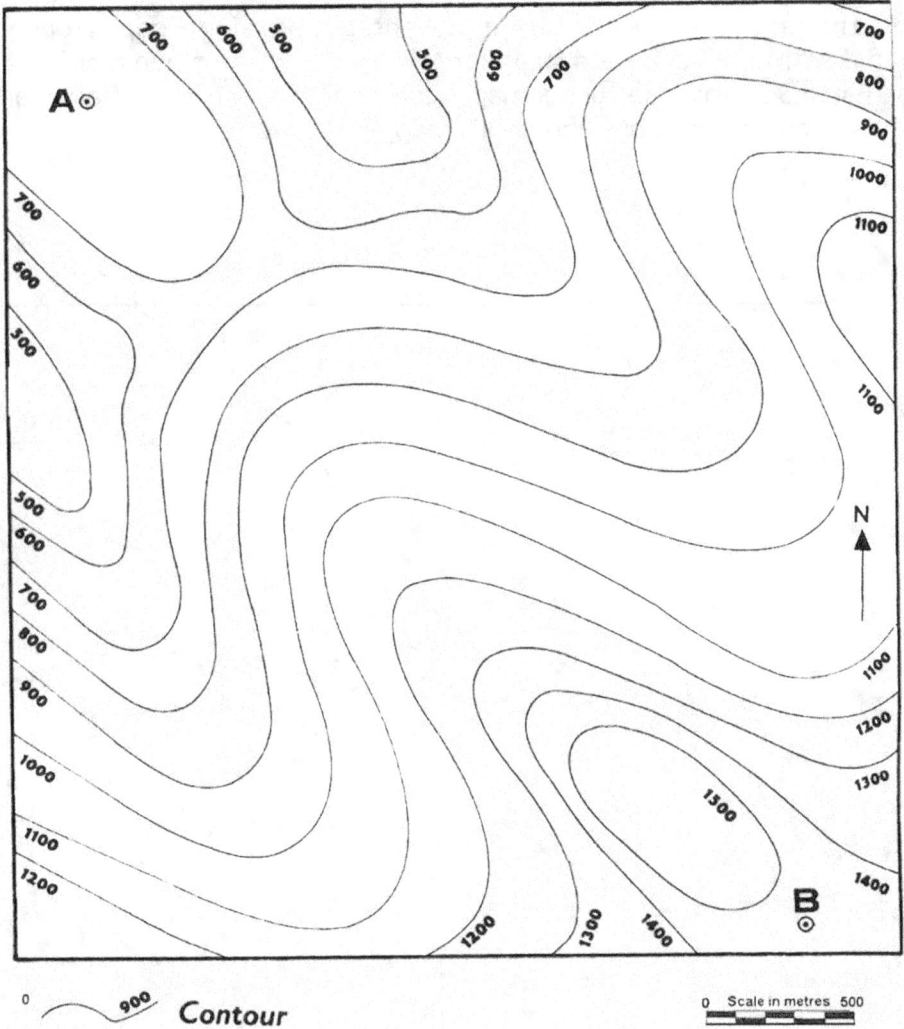

Draw a topographical cross-section between SPOT HEIGHTS A and B using the scales as shown on the map. Draw the cross-section in a box giving sufficient depth so that the cross-section can be fully shown and some additional depth is given below the minimum height above sea level.

EXPERIMENT 1.3 continued

QUESTIONS:

1. What feature lies between spot heights A and B?
2. When drawing this cross-section, what allowances to reality must the drawer make?
3. What is the GRADIENT of the steepest slope along section A-B?
4. If this is open grassland country and in good weather, what would be an estimate of the time it would take to walk (unburdened) from A to B?

CONCLUSIONS:

1. What would be the (a) advantages and (b) disadvantages of using a paper topographical map over that of an electronic, hand-held Mapping Ap. or computer screen map?
2. Why would an Earth Scientist use such a map? (consider pre- and post- field work activities)
3. What are some general conclusions about this mapping exercise?

Chapter 2: Rock-forming Minerals

EXPERIMENT 2.1   One to three Lessons

SOME COMMON ROCK-FORMING MINERALS

AIM: To describe, draw and then identify some common minerals from which rocks form.

MATERIALS: Mineral specimens (quartz; orthoclase; plagioclase; calcite; hornblende; olivine in basalt rock; biotite; muscovite) hand lens, streak plates, and Mohs' scale of hardness kit.

BACKGROUND:

Minerals are pure, inorganic chemical compounds having definite properties. These properties which are useful in identifying specimens include:-

1. COLOUR - the frequencies of light coming from the specimen when viewed in normal white light. Simple colours (e g black, red-brown etc.) are used to describe this property. A mineral may have a variety of colours.

2. STREAK - the colour of the powdered mineral. If soft, the specimen may be quickly drawn across a STREAK PLATE (an unglazed tile -white for coloured minerals, black for white or clear minerals). If the specimen is hard, it may be scratched with a knife or a harder mineral and the colour of the scratch observed.

3. CRYSTAL FORM (Habit) – the overall appearance of the specimen in the way that it has been formed e.g. CLEAVAGE BLOCK (individual crystals not seen, but many flat crystal surfaces observed giving a blocky appearance); BOTRYOIDAL (many rounded lumps together); FIBROUS (long strands or fibres); RADIATING (long crystals radiating outwards from a common point); GRANULAR (grouped like grains of sand); AMYGDALOIDAL (crystals growing in holes in rock); FOLIATED (as sheets); PISOLITIC (pea-sized spheres); or MASSIVE (no crystals seen).

4. LUSTRE - the way light reflects off the specimen's surface. It may vary from surface to surface (i.e. a specimen may have a range of lustre). Lustre may be:

         METALLIC -hard, shiny like a metal
         SHINY – very reflective like gloss paint
         VITREOUS -glassy, with some depth
         PEARLY -like a pearl or button
         SILKY -shiny with a fibrous look
         GREASY -oily appearance like soap
         RESINOUS -with a dull transparent look
         DULL -very little shine

EXPERIMENT 2.1 Continued

5. CLEAVAGE - the ability to break naturally along flat surfaces. The best test is to actually cleave the mineral but this is often not desirable so one has to look for the number of cleavage planes which may meet together. Minerals may have:

NO CLEAVAGE - but the mineral may have crystal faces e.g. quartz or cleavage in:

ONE DIRECTION called basal cleavage which gives sheets
TWO DIRECTIONS giving lines of small steps
THREE DIRECTIONS giving points or corners (which also may be cut off) or even in
FOUR DIRECTIONS giving four-cornered points like a pyramid (rare).

6. HARDNESS - resistance of the mineral to be scratched. Usually compared to a standard set of minerals (MOHS' SCALE) or a set of material for approximations (FIELD SCALE):

MOHS' SCALE	FIELD SCALE
1. TALC (softest)	Thumb nail 2.5
2. GYPSUM	Coin 3.5
3. CALCITE	Glass 5.0 -6.0
4. FLUORITE	Knife 5.5 -6.0
5. APATITE	File 6.5- 7.0
6. ORTHOCLASE	
7. QUARTZ	
8. TOPAZ	
9. CORUNDUM (ruby)	
10. DIAMOND	

7. SPECIFIC GRAVITY - the density of the mineral compared to that of water i.e. how many times the mineral in heavier than an equal volume of water (density of water = 1g/cc). Heft is a word to generally describe the mineral's heaviness in vague terms e.g. heavy, medium, light).

8. OTHER PROPERTIES - any other unique observation e.g. flexible, fluorescent, radioactive, magnetic, chemical reaction, taste (careful!) etc.

9. SKETCH:

A sketch is to be drawn for selected minerals which clearly show some of the above features. These sketches should be to scale and in appropriate colour. This is done by:

EXPERIMENT 2.1 Continued

    a. selecting an appropriate ORIENTATION (side) from which the sketch is made;

    b. studying the OUTLINE and then drawing it lightly to scale so that about two or three sketches will fit per page (you may put a border around each if you wish);

    c. re-examination of the specimen (possible with a hand lens) to determine the main INTERNAL FEATURES for exaggeration and drawing these within the outline. If there is considerable detail, only draw a representative part of it; and

    d. once the sketch is complete, go over the outline and main features with heavier pencil (or even black ink if you are careful).

PROCEDURE:

1. Draw up a **TABLE** (with a full page turned sideways) under DATA & OBSERVATIONS using the following headings:

    CODE   COLOUR   STREAK   LUSTRE   CLEAVAGE   HARDNESS   OTHER

2. Test and describe each specimen for each of the properties (given above).

3. DRAW each specimen TO SCALE (e.g. x2 or x3 etc.) and **IN COLOUR** where appropriate.

4. Use the text, display minerals, computer and the KEY (provided below) to NAME the specimen. WARNING: not all the minerals may be in the list below!

DATA and OBSERVATIONS: (Table and Sketches as appropriate)

CONCLUSIONS:

1. LIST the code numbers of the specimen, the specimen's names and at least two DISTINCTIVE properties of each mineral which will then help in the quick identification of that mineral.

2. COMMENT on the use of mineral properties (advantages/disadvantages) in identifying unknown minerals.

3. What are some likely errors which may limit correct identification of minerals in the field?

EXPERIMENT 2.2	One lesson to set up then several days

GROWING CRYSTALS

AIM: To grow well-formed crystals from solution.

MATERIALS: Copper Sulfate or Alum or Chrome Alum; petri dishes; de-ionized water; Salicylic Acid crystals; beakers (250 ml); sticks; thread; large test-tube & rack; Bunsen burner or hotplates.

BACKGROUND: Crystals can form from evaporation of water from a mineral solution or by cooling a hot solution down until the solute "un-dissolves" and precipitates as crystals. Once a small crystal has formed and is still in a saturated solution, it may continue to grow as more ions add to it from the solution (accretion).

PROCEDURE:

PART A: CRYSTALS FROM HOT SOLUTIONS

Care! Heating water in a test-tube can be dangerous if done too quickly. Use paper to hold the tube, point it away from others and heat slowly on a low flame.

1. Add a small amount (about half a spoonful) of Salicylic Acid (a solid organic acid used in making Aspirin) to about half a **large** test-tube.

2. CAREFULLY boil the water and dissolve the solid.

3. When completely dissolved, hold the outside of the test-tube under cold water and observe carefully.

QUESTIONS:

1. What happens as the solution cools? Why?

2. Describe any shapes. Sketch the apparatus and observations in the Results section.

PART B: GROWING LARGER CRYSTALS

1. Dissolve Copper Sulfate crystals (or Alum or Chrome Alum) in a little hot water - enough to fill a small petri dish.

2. Pour the solution in the dish and leave overnight in a forced-air fume hood or place of good air circulation.

3. Examine the shapes and any other feature using a hand lens or microscope.

Leave about half a page under RESULTS for a sketch and description of the results next day (it may take more than one – why?)

EXPERIMENT 2.2 Continued

4. Select the BEST SHAPED crystal and secure it with a loop of cotton thread, so that when attached to the stick it can hang in the beaker about 1-2 cm. from the bottom (see diagram next page).

5. Cover the beaker with paper (why?) and leave overnight.

6. Each day thereafter (for about a week), top up the solution (it MUST always be saturated. Why?) and check the SIZE and SHAPE of the crystal. Remove any imperfections or smaller crystals on the string.

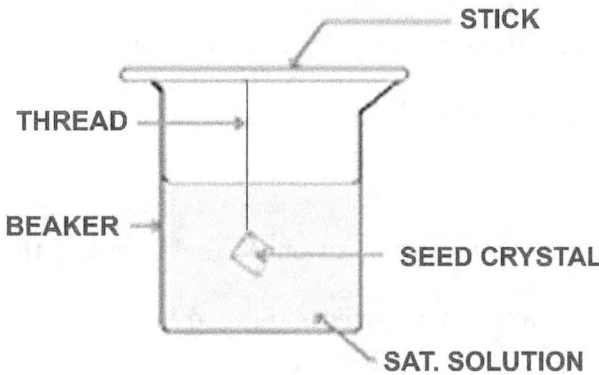

7. RECORD the size (e.g. width) of the crystal each day (in millimetres) and enter the data in a table in RESULTS.

CONCLUSIONS:

1. Comment generally on how these crystals can grow from solution and how smaller crystals grow to bigger size.

2. What are the limitations to the size of the crystals?

3. Is there any indication within the crystal as to how they grow from smaller crystals?

4. Where in nature do crystals grow in this way (Internet research needed).

EXPERIMENT 2.3　　　　　　　　　　　　　　　　　　　　　　　　　　**One lesson**

MEASURING SPECIFIC GRAVITY

AIM: To determine the specific gravity of a mineral.

MATERIALS: mineral specimen; electronic balance; large beaker (250 ml); thread

BACKGROUND: SPECIFIC GRAVITY is the density of the mineral compared to the density of water (= 1 gram/cubic centimetre   Note: 1 cc = 1 millilitre approx.)

$$\text{i.e. SG} = \frac{\text{Density of Mineral}}{\text{Density of Water}}$$

(note: there is no unit of measurement – it is a ratio)

and DENSITY = Mass (in grams)/Volume (in ml)

PROCEDURE:

1. Weigh each specimen carefully in air on the balance and the note the value.

2. Half-fill a 250 ml beaker with water and place it on the balance which was been zeroed (TARED).

3. Secure the specimen by a loop of thread and completely submerge it below the water in the beaker. The reading on the balance represents the weight of water displaced by the specimen and thus its **volume** (Archimedes' Principle).

4. Repeat the measurements several times and take an average.

5. Using the mass and this value for volume, calculate the density of the mineral. Since S.G. is a comparison to the density of water which is 1 g/cc then this numerical value calculated is the S.G. of the mineral.

6. Using the average value of the S.G. and the table (given at the end), identify the white mineral

7. Record all data and any observations Also give the ERROR of any measurement

REMEMBER that for any ANALOGUE instrument (where all divisions can be seen), error = half a unit for any DIGITAL instrument (only one unit can be seen) = one unit

When calculations are made using errors, it is best to use PERCENTAGE ERRORS which are then ADDED to give the overall error for the calculated answer.

e.g. % error for mass   = $\dfrac{\text{instrument error}}{\text{measured value}} \times 100$

To be more accurate, the measurements can be taken several times and calculate the S.G.

EXPERIMENT 2.3 Continued

CALCULATIONS: Use the formula given in BACKGROUND to calculate S.G. and give an overall % error

CONCLUSIONS:

1. What was the calculated value for S.G. and its error?
2. From a list of S.G. what is this mineral?
3. How does the experimental value agree with the book value for this mineral?
4. Comment on any student, environmental or instrument errors.
5. How could this experiment be improved?

TABLE OF SOME SPECIFIC GRAVITIES

Mineral	S.G.
Borax	1.7
Halite	2.1
Calcite	2.7
Plagioclase	2.6
Aragonite	2.9
Strontianite	3.8
Barite	4.5
Anglesite	6.3
Cassiterite	7.0

EXPERIMENT 2.4 *Two lessons*

GEOCHEMISTRY

AIM: To observe and describe some common tests for metal ions and some salts and hence determine the chemistry of a given unknown mineral.

MATERIALS: Bunsen burner & ceramic mat; paper clips; wooden tongs; test-tubes & racks; beaker; limewater solution; safety equipment (apron, glasses, gloves); powdered metal salts (e.g. copper sulfate, sodium chloride, barium carbonate, strontium nitrate, Lithium chloride) ;powdered minerals e.g. sulfides (galena); carbonates (dolomite, calcite, magnesite); chloride (halite); and sulfate (epsomite); 2M Hydrochloric acid in dropper bottle; dropper bottles of Barium Nitrate solution and Silver Nitrate solution.

BACKGROUND:

Chemical testing of various minerals has been on-going for a long time. At the end of the 19th Century, German chemists (notably Bunsen and Kirchhoff) were using the newly discovered spectroscope to analyse the light coming from coloured flames produced by burning metal salts in a gas flame. This later led to the development of the Flame Spectrometer which is used today to identify the common metals in minerals. Sometimes the mineral specimens are too small to identify by physical properties, so chemical testing is required.

Several ion groups can also be tested in the school laboratory. CARBONATES give odourless CARBON DIOXIDE gas in acid and SULFIDES give smelly HYDROGEN SULFIDE (very poisonous! Do not smell too much of it!) with acid. CHLORIDES give a white precipitate (of Silver Chloride) with Silver Nitrate solution (stains the skin!) and <u>soluble</u> SULPHATES give a white precipitate (of Barium Sulfate) with Barium Nitrate.

PROCEDURE:

PART A: FLAME TESTS (one lesson)

1. Set up a Bunsen burner so that it sits on a protective ceramic mat.

2. Unfold a paper clip to a straight wire and, holding one end in wooden tongs, heat the other end in the BLUE flame of the Bunsen burner.

3. Dip the hot end of the wire into water in a beaker and then dip it into a sample of a metal salt to pick up ONE or TWO small crystals (important use the smallest possible!).

4. Heat the crystal(s) in the BLUE flame of the Bunsen burner and note the colour of the flame.

5. Repeat parts 3 and 4 (above) for each of the selection of metal salts. Remember to CLEAN the end of the wire in water after each test and note the colour of the flames for the metal salts (copper sulfate, sodium chloride, barium carbonate, strontium nitrate, lithium chloride).

EXPERIMENT 2.4 continued

PART B. CHEMICAL TESTS (Lesson 2)

TEST FOR CARBONATES

1. Into separate test tubes place a small amount (half-pea size) of each of the metal carbonates provided (dolomite, calcite, magnesite). Remember the positions and names of each sample.

2. Just cover each sample with a little acid (CARE! It is caustic and will damage tissue) and observe any reaction (cautiously sniff any gas, waft a little towards the nose using your hand in a wave like action....there may also be some pungent acid vapour).

3. Place a glass rod which has been dipped into some limewater and note any changes.

TEST FOR CHLORIDES

1. Into a test tube place a very small amount of Halite (Sodium Chloride).

2. Cover with about half a centimeter of water and shake to dissolve some of the mineral.

3. Add two drops of Silver Nitrate solution (CARE!) and observe.

4. Note any reaction.

TEST FOR SULFATES

1. Into a test tube place a very small amount (2-3 rice grains) of Epsomite (Magnesium Sulfate).

2. Cover with about half a centimeter of water and shake to dissolve some of the mineral.

3. Add two drops of Silver Barium Nitrate solution and observe.

4. Note any reaction.

TEST FOR SULFIDES

(Students are warned not to make this gas in large amounts and to sniff it very cautiously. IMMEDIATELY after all students in the group have made an observation, the test tube and its contents must be handed in for removal to the fume hood)

1. Into separate test tubes place a very small amount (about the size of a rice grain) of the lead sulfide (e.g. galena).

2. Cover with about half a centimetre of acid and observe. CAUTIOUSLY smell any gas given off by waving a hand across the top of the test-tube towards the nose.

3. Hand in the test-tubes for placement into a fume hood (or outdoors) for disposal in a bucket filled with water.

EXPERIMENT 2.4 continued

RESULTS:

1. Record all of your observations in a TABLE;
2. Find out the CHEMICAL NAME of each mineral tested and write a WORD EQUATION for each reaction;
3. DRAW a representative sketch (one half to one page in two dimensions with labels) of any one of the test-tube reactions.

CONCLUSIONS:

Use the following questions to write a detailed conclusion about this activity:-

1. What was the TEST and it's RESULT for each METAL ION (flame test) and for each NON-METAL GROUP (carbonates etc.)? LIST or table summary.

2. Why was de-ionized water used for making up solutions and dissolving the minerals?

3. What would be the main ERRORS in these tests (be specific for Parts A & B)?

4. In what situations may a geologist in the field use some of the tests in Part B? EXPLAIN.

RESEARCH: (Optional)

What are some of the instruments and methods used by Geochemists to analyse specimens to find their composition.

Chapter 3: Igneous Rocks - The Beginning

EXPERIMENT 3.1	One or two lessons

SOME COMMON IGNEOUS ROCKS

AIM: To examine and describe some common igneous rocks.

MATERIALS: Hand Lens, Rock specimens of Granite, Gabbro, Andesite, Trachyte, Diorite, Rhyolite, Basalt, a Porphyry, Pumice, Tuff and a Volcanic Glass.

PROCEDURE:

1. Examine each specimen in turn using the hand lens.

2. Describe each rock in turn using the following descriptors and a table to record the data in Results:

COLOUR: overall colour appearance e.g. pink with black spots, grey, black etc.

CRYSTAL SIZE: as coarse (> 2mm), fine (<2 mm), glassy (none) or porphyritic (has bigger crystals in a background of smaller crystals).

TEXTURE: is the way the crystals appear to be locked together e.g.

    PHANERITIC – all crystals large (> 2mm) and interlocked
    APHANITIC – crystals mostly not visible to the naked eye but can be seen under a microscope.
    PORPHYRITIC – big crystals (called "Phenocrysts") in a smaller crystalline background (or "Matrix"). Indicates two stages of cooling.
    GLASSY (HOLOHYALINE) no crystals seen, not even under a microscope.
    PYROCLASTIC broken and angular fragments formed by volcanic explosion.

COMPOSITION: if you can see crystals, NAME them using simple common terms:
    e.g. quartz looks glassy grey
        orthoclase is pink and blocky
        plagioclase is white or shiny blue gray – may be long and shiny in dark rocks
        biotite is black and very shiny flakes
        olivine is dark green and glassy granules like sugar
        hornblende is black, thin and long.

    If no crystals seen (often = very fast cooling) then write "none seen"

STRUCTURES: what shapes can be seen within the rock (mostly in Extrusive rocks) e.g.
    MASSIVE – no structure, only uniform crystals or no crystals at all
    VESICULAR – gas bubbles
    AMYGDALOIDAL – gas bubbles filled with mineral.
    FLUIDAL flow lines (may look like part layers)

EXPERIMENT 3.1 Continued

SKETCH - draw three sketches of the specimen which show:
  Phaneritic Texture with large crystals, mostly equal sizes. Name and label the main minerals seen in this specimen;
  Porphyritic Texture of large crystals within a background of smaller crystals. Name and label the larger crystals (Phenocrysts); and
  Holohyaline Texture with no crystals but showing structures such as flow lines.

DATA and OBSERVATIONS: Construct a table of data from descriptions and sketches to scale in colour. Sketches should be of a suitable size e.g. two or three to a page.

CODE	COLOUR	TEXTURE	MINERALOGY	GRAIN SIZE (each mineral)	STRUCTURE (if seen)

CONCLUSIONS:

1. In your Conclusion, try to name each of each specimen and for each, give a few KEY WORDS to remember each rock e.g. Pumice could be "grey, bubbles"

2. Comment on the SIGNIFICANCE of each rock. i.e. HOW (and WHERE) it might have been formed. That is, CLASSIFY each rock as either:

  INTRUSIVE (formed below surface) or EXTRUSIVE (formed on surface) and
  FELSIC light in colour) or MAFIC (dark in colour) e.g. "Rock No. # is intrusive, mafic.

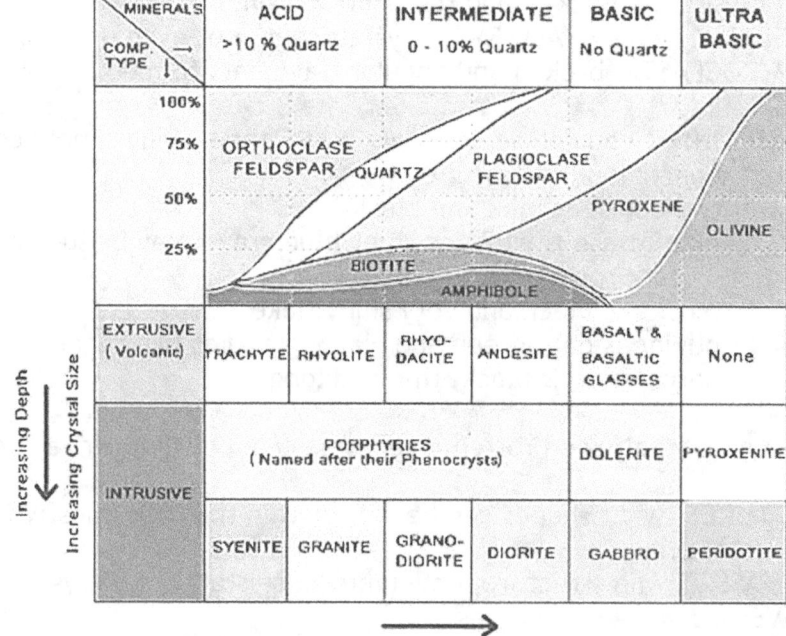

Chapter 4: Sedimentary Rocks

EXPERIMENT 4.1: One Lesson

SOME COMMON SEDIMENTARY ROCKS

AIM: To examine and describe some common sedimentary rocks.

MATERIALS: Bottles of sediment and water, collection of sedimentary rocks (e.g. conglomerate, sandstone, shale, mudstone, greywacke, coal and limestone), hand lenses, watch or clock with second hand. Small dropper bottles of dilute acid.

PROCEDURE:

PART A: FORMATION OF LAYERS FROM SEDIMENTATION

1. Closely examine the bottles of sediment and water provided (DO NOT remove the tops). Ensuring that the top is secure, shake the bottle firmly to mix all of the sediment. What does this represent in nature?

2. Stop shaking the bottle and stand it upright on the table. As soon as you stop shaking it, time how long for each type of sediment to stop falling. The biggest sediment is GRAVEL, the next is SAND, the third is SILT and the rest (which may stay **suspended** in the water for a long time) is CLAY (this might take several **days** to settle!).

3. Record these times in a table in the Results (Part A).

How have the sediments settled (in what order)? Record the answer in the Results and DRAW A SKETCH of the sediments in the bottle (a two-dimensional side view).

PART B: DESCRIPTION OF SPECIMENS OF SEDIMENTARY ROCKS

4. Closely examine each rock specimen and describe your observations in a table in Results (Part B). Use the following descriptors:

> **COLOUR** general overall colour
> **GRAIN SIZE** as **coarse** (>2mm), **medium** (1mm - 2mm) **fine** (<1mm)
> or **none** (see the table of the WENTWORTH SCALE provided)

Wentworth Scale

GROUND	SIZES	TYPICAL ROCK
Boulders	Greater than 256 mm	Conglomerate (rounded) or Sed. Breccia (angular)
Cobbles	64 - 256 mm	
Pebbles	4 - 64 mm	
Granules	2 - 4 mm	
Very coarse sand	1 - 2 mm	Sandstone
Coarse sand	0.5 - 1mm	
Medium Sand	0.25 - 0.5 mm	
Fine Sand	0.125 - 0.25 mm	
Very Fine Sand	0.062 - 0.125 mm	
Silt	Less than 0.062 mm Looks smooth Feels gritty	Siltstone

EXPERIMENT 4.1 continued

GRAIN SHAPES will often indicate the amount of transportation which the grains have undertaken as well as their hardness. Use the following diagram:

SHAPE - will vary from VERY ANGULAR to WELL-ROUNDED

| VERY ANGULAR | ANGULAR | SUB-ROUNDED | ROUNDED | WELL-ROUNDED |

SORTING shows the arrangement of the grains as:-

WELL-SORTED – all grains about the same size
MEDIUM SORTED – most grains about the same size
POORLY SORTED – most grains of different sizes

COMPOSITION - the nature of the rock's GRAINS (clasts or particles), MATRIX (particles between the grains) and CEMENT. Try to estimate the relative PERCENTAGE of the composition of the grains. Grains (and matrix) are usually QUARTZ, FELDSPAR or ROCK FRAGMENTS (rock fragments give LITHIC sandstones, conglomerates etc.). CEMENTS are usually CALCITE (bubbles with acid), SILICA (i.e. quartz which gives a hard rock) or IRON OXIDES (gives a brown or yellow colour)

STRUCTURES any layers, fossils, bigger grains etc. The most common structures seen are:
  LAMINA or fine layers
  GRADED BEDDING – big clasts ranging (upwards) to smaller clasts but usually only seen in larger scale
  MASSIVE – no structures, all uniform appearance
  FOSSILS – plants indicate freshwater, shells and corals are marine.

and **SOURCE (or PROVENANCE)** an idea of the ENVIRONMENT which formed the sediment e.g.
  FINE (SILT or MUD) = still water (lake, deep ocean)
  SAND = medium flow (river or ocean current near shore)
  BIG PEBBLES = fast mountain stream or storm beach
  BIG, ANGULAR = an AGGLOMERATE from landslides or glacier
  LIMESTONE = usually a coral reef and may have fossils

Use the textbook or other sources to identify each rock. Write the number of the specimen, the rock's name and then its classification (either CLASTIC or NON-CLASTIC) and perhaps two key words for later identification.

EXPERIMENT 4.1 continued

DATA and OBSERVATIONS:

PART A: Describe the rate of settling and draw a table for each fraction (gravel/sand/mud) showing rate of settling against time. Sketch the final layers in the bottle.

PART B: Table of rock properties for each specimen and selected sketches. Copy and complete:

Specimen Number	Overall Colour	Texture (Grains)			Composition	Structures	source	Name
		Size	Shape	Sorting				

Do a **SKETCH** of conglomerate <u>and</u> either shale **or** limestone (show any fossils present).

CONCLUSIONS:

In your conclusion for **Part A**, EXPLAIN why the sediments have settled like this and comment on the time taken for each. Comment on the factors which would control the speed and order of settling

For **Part B**, LIST the names of the rocks and write two KEY WORDS to identify the rock and also give a CLASSIFICATION for each as CLASTIC (particles can be seen) or NON-CLASTIC (no particles e.g. biological or chemical).

Also comment on any internal structures or features seen in each of the rocks and how they and the rock's composition relate to their environment of formation.

EXPERIMENT 4.2   One Lesson

MODAL ANALYSIS OF SEDIMENT

AIM: To use the technique of Modal Analysis to deduce the nature of a sediment

MATERIALS: Sediment mixture (silt, fine sand, coarse sand, fine gravel), sieve sets, 500 ml plastic measuring beakers, 100 ml measuring cylinders, hand lenses.

BACKGROUND: The nature of the grains in a soil (or Sedimentary Rock) can indicate the conditions under which the soil (or rock) was formed.

Column graphs of **size** against **percentage of volume** will suggest the most common (**mode**) of size of grain and hence the mode of formation. **Sorting** is shown by the relative percentages graphed for each size.

Sedimentary Rocks must first be gently agitated or have their cements dissolved (in acid) before their grains can be tested.

PROCEDURE:

1. Arrange the sieves in stacks with the smallest size on the bottom and the largest on top.

2. In turn, pour 500 ml of the sediment into the top of the sieve set, cover and then shake the combined set vigorously for a few minutes.

3. Separate the sieve set into its separate sieves and measure the volume of each **fraction** (amount) in each sieve.

4. Convert each fraction value to a **percentage volume** of the original volume (e.g. 500 ml) and graph these against the **sieve sizes** as a column graph e.g.

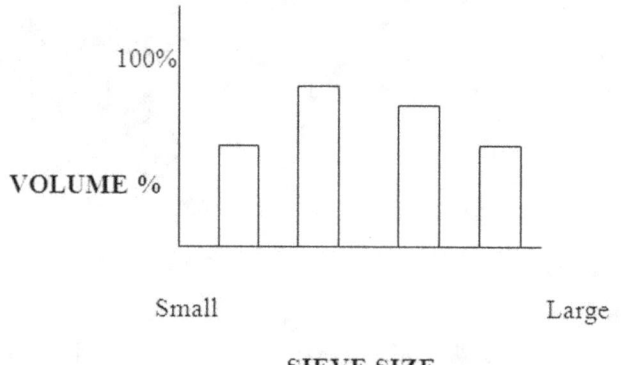

EXPERIMENT 4.2 continued

DATA and OBSERVATIONS:

Draw a full page column graph (as above) which accurately reflects the measurements taken for each sediment sample. Clearly label each with the location of the sand sample.

CONCLUSIONS:

1. Comment on the sorting for each sand sample;

2. Suggest reasons for any differences or similarities between:
   a) Beach and Dune Sands
   b) River Gravel and River Sand
   c) Beach Sand and River Sand

3. What errors might occur in using VOLUME as a measurement?

4. Can you suggest a more accurate measurement?

5. Comment on the general accuracy and usefulness of this experiment in the study of sedimentary rocks.

EXPERIMENT 4.3	One or two lessons

POROSITY AND PERMEABILITY

<u>Aim</u>: To measure and compare the porosity and permeability of some common sediments.

MATERIALS: Sand, silt & gravel, 250 ml measuring beakers, 100 ml measuring cylinders, large filter funnels, filter paper, stop watches or clock with a second hand.

BACKGROUND:

Hydrologists (water Scientists), Agriculturalists, Agronomists (Soil Scientists) need to know the porosity and permeability of soils and sedimentary rocks (**WHY?**). **POROSITY** is the amount of pore space (open space) between the grains of sediment. It is found by:

$$\text{POROSITY \%} = \frac{\text{Volume of pore space}}{\text{Total volume}} \times \frac{100}{1}$$

PERMEABILITY is the rate at which water moves through sediment and this depends upon several factors including the nature of the sediment, the cross-sectional area of the layer of rock and the height of water intake over that of the outflow (i.e. the Head of water).

PROCEDURE:

(a) POROSITY

1. Fill the measuring beaker to the 250ml level with sand or silt or gravel and check that the surface is level. This is the **total volume** of the soil.

2. Measure exactly 100ml of water in the measuring cylinder and slowly pour it into the beaker until the water level JUST reaches the soil surface. Record how much water was poured into the beaker. A second 100ml measurement may be needed with some soils so keep a record of the total amount of water poured into the beaker.

3. Determine the total amount of water poured into the beaker. As the water will fill the pore spaces, this amount is the volume of pore space.

4. Calculate the porosity as a percentage for this sediment.

5. Carefully empty this wet sediment into the appropriate waste bucket (one for EACH of the sediments) and repeat this method for the other two sediments.

(b) PERMEABILITY

1. Place a small wad of torn filter paper into the bottom of the filter funnel.

2. Measure 100 ml of water in the measuring cylinder.

3. Place the filter funnel into the top of an empty container (e.g. another measuring cylinder) and, starting from a set zero time, pour as much water as possible into the funnel (and maintain the level by constant pouring) and time how long it takes for all of the water to flow through the funnel. This is a CONTROL EXPERIMENT which will give the "zero error" due to the resistance of the wad of paper.

EXPERIMENT 4.3 continued

4. Repeat this experiment with each of the three sediments by placing 100 ml of DRY sediment into the filter funnel (with its wad of paper) and then pouring in 100 ml of water from the beaker, measuring the time for this to flow through.

5. Repeat this part with each sediment WET (i.e. do it with dry, then wet sediment and measure times).

6. DRAW a labeled sketch of this apparatus e.g.

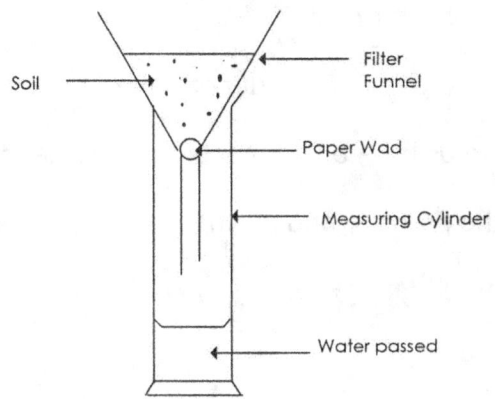

DATA and OBSERVATIONS:

a) Porosity -

Soil	Total Volume of Soil	Volume of Water Added	Percent Porosity
Silt			
Sand			
Gravel			

Calculations:

b) Permeability -

Control experiment (no soil):
Time:
Volume passed:

Main Experiment

	Silt		Sand		Gravel	
	Dry	Wet	Dry	Wet	Dry	Wet
Time						
Volume passed						

EXPERIMENT 4.3 continued

CONCLUSIONS:

1. LIST the sediments in ORDER from BEST to LEAST for (a) POROSITY and (b) PERMEABILITY.

2. Are these lists the same? Explain any differences or similarities.

3. What was the purpose of the CONTROL EXPERIMENT?

4. Which of the permeability methods (DRY or WET) would be a more honest measure of a sediments true permeability? Explain.

5. What could change the porosity of a sediment in nature?

6. When can a rock be porous but not permeable (impermeable)? Give an example.

7. Why is knowledge of the porosity and permeability important in using natural resources?

Chapter 5: Metamorphic Rocks

EXPERIMENT 5.1   One Lesson

SOME COMMON METAMORPHIC ROCKS

AIM: To examine and describe some common Metamorphic Rocks

MATERIALS: Hand lenses, set of metamorphic rocks (gneiss, hornfels, phyllite, greenstone, slate, marble and schist), and dropper bottles of dilute acid.

PROCEDURE:

1. Carefully examine each specimen with a hand lens.

2. In a table in Results, describe each rock under the following terms:

    COLOUR overall colour

    TEXTURE is it FOLIATED or NONFOLIATED?

    COMPOSITION if possible describe some minerals seen (revise the practical work on minerals)

    (additional experiment : with care, test the non-foliated rocks with dilute acid. What does this show about the one that reacted?)

    OTHER any other description worth noting.

3. In the Results, SKETCH the GNEISS to show its texture.

DATA and OBSERVATIONS:
Draw up a Table of Properties and make sketches of those rocks which show interesting colour or textures e.g.

Rock Number	Overall Colour	Texture	Composition (Minerals)	Other	Name of Rock

CONCLUSIONS:

In your Conclusion, LIST the names of each mineral, give two or three KEY WORDS to help you remember the rock AND give the PARENT ROCK from which the specimens came by metamorphic action.

CLASSIFICATION OF METAMORPHIC ROCKS

TEXTURE				ROCK NAME	COMPOSITION	PARENT ROCK	METAMORPHIC PROCESS
	Foliated	Schistose	Very fine-grained	SLATE	Abundance of dark, flaky and/or prismatic silicae minerals (micas, chlorite, talc, serpentine, hornblende, etc.); quartz	Shale; tuff	INCREASE REGIONAL ↓
			Fine-grained	PHYLLITE		Shale; tuff	
			Medium- to coarse-grained	SCHIST (var. Mica schist, chlorite schist, amphibole schist, etc.)		Shale; intermediate to mafic igneous rocks	
		Gneissic	Medium- to coarse-grained	GNEISS (var. Garnet gneiss, granite gneis, etc.)	Feldspar abundant; varying amounts of quartz and dark silicate minerals (such as amphiboles, pyroxenes, micas, and garnet)	Felsic to intermediate igneous rocks; arkose; graywacke; mica schist	
	Nonfoliated	Granoblastic	Medium- to coarse-grained	METAQUARTZITE	Quartz greatly predominant	Normal and quartzose sandstones	Contact
				MARBLE	Calcite and/or dolomite, with or without Ca-Mg silicates	Limestone or dolomite, with or without impurities	Contact
		Hornfelsic	Fine- to very fine-grained	HORNFELS	Dark silicate minerals predominant	Shale; slate; intermediate to mafic extrusive rocks	Contact
				ANTHRACITE	92 - 98% carbon	Peat, lignite, coal	Contact

Chapter 6: Weathering and Erosion

| Experiment 6.1 | One Lesson |

WEATHERING

AIM: To examine and describe causes and features of weathering.

MATERIALS: Limewater, test-tubes & rack, dropper bottles of dilute acid (2M HCl), watch glasses, pieces of limestone, fresh & weathered granite, hand lenses, metal tongs, Bunsen burners and mats

BACKGROUND

Weathering can be PHYSICAL or CHEMICAL. Both cause the breakdown of minerals and rocks; chemical weathering by chemical change and physical weathering by breakdown into smaller pieces.

Limewater (Calcium Hydroxide solution) is a test solution for carbon dioxide gas. It reacts slowly with CO_2 to form a faint white cloud of calcium carbonate. **CAUTION:** Limewater is also slightly caustic (corrosive to skin).

All **carbonates**, such as calcium carbonate, react with acid to produce CO_2 (and the salt of that acid, e.g. chloride if hydrochloric acid is used)

(in advance: DEMONSTRATION- pour a little limewater into a watch glass and leave .)

PROCEDURE:

PART A: Weathering of Granite (overall)

1. Observe pieces of fresh and weathered granite. Feel both specimen and comment on their overall differences.

2. Carefully examine a FRESH piece of granite and identify by sight the main minerals in it (revision from the Igneous Rocks Experiment).

3. Using the hand lens, carefully examine a weathered piece of granite. Try to locate the original minerals (may be changed now) by their relative size, shape and abundance). Describe (colour, hardness, other features) how these minerals have changed.

4. Set out these observations in a Table and SKETCH and label the <u>weathered</u> granite to a good scale (e.g. half page) and in colour.

PART B: Physical Weathering of Granite

1. WEAR EYE PROTECTION. Carefully heat a small sample of fresh granite by holding it with metal tongs in a blue flame of a Bunsen burner.

2. Quickly drop the hot rock into a glass test-tube of cold water (these two steps may have to be repeated several times). Describe any changes observed.

EXPERIMENT 6.1 continued

PROCEDURE: continued

PART C: Chemical Weathering of Limestone or Marble

1. Place a few pieces of limestone or marble rock into a test tube and CAREFULLY add enough acid to JUST COVER the solid.

2. Test any gas with a drop of limewater hanging on the end of a bacterial loop. What happens? What does this prove?

3. SKETCH the main apparatus.

DATA and OBSERVATIONS:

Give descriptions of observations for each part with special attention to the differences between fresh and weathered specimens (e.g. in granite, indicate the minerals and how they have changed). Include WORD EQUATIONS for any chemical changes observed (if possible) and draw neat, well-labeled sketches as appropriate.

CONCLUSIONS:

1. Comment on what was learnt in each of the three sections about weathering.

2. Give WORD EQUATIONS for any chemical changes (e.g. minerals in granite, acid on Limestone etc.).

3. Relate the artificial, laboratory actions (e.g. using Bunsen burners, adding acid etc.) to the real source of change for the actions observed in this experiment.

4. Discuss any errors which may have prevented the appropriate observations from being made. Suggest any improvements.

EXPERIMENT 6.2   One Lesson

INTRODUCTION TO STREAM TABLE EXPERIMENTS

AIM: To examine how sediment can be eroded, carried and deposited by water.

MATERIALS: Long white trays, small deep trays, retort stands (rod), boss heads and clamps, 500 ml beakers, sand

BACKGROUND:

Sediment is easily eroded by running water and is carried downslope to where it can be deposited. In doing so, the sediment often forms features typical of stream patterns.

PROCEDURE:

1. Set up the following apparatus:

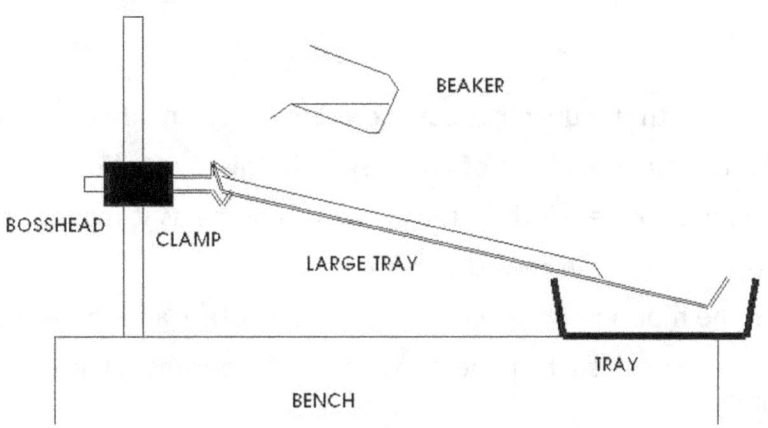

2. Set the tray in an horizontal (level) position (the base of the Bosshead is about 17 cm. above the top of the desk).

3. Add about four beakers full of sand to the tray and moisten until the sand is wet but hard and easy to mold.

4. Smooth the sand out as a flat layer which takes up most of the tray (as shown above).

5. Adjust the Bosshead until the tray is at an angle of about 5 degrees (Bosshead base is about 21 cm from the bench) so that the flat tray slopes and runs into the smaller tray.

6. Calculate the slope of this surface (change in vertical height/change in length).

7. Using the beaker full of water, gently and consistently pour a continual stream of water onto the sand (as shown in the diagram) from about an height of 10 cm.

8. Make general observations as to: shape of the channel; speed of water; and other general effects on the sand.

EXPERIMENT 6.2 continued

PROCEDURE: continued

9. Taking a LINE OF SIGHT (i.e. imagine a straight line from where the water was poured onto the sand and where it flowed out at the end) along the sand surface, measure (in cm.) the MAXIMUM WIDTH of the valley formed by the stream (several may have to be taken and then an average made).

10. Drain off the water into the collecting tray (or a waste bucket if this is full...NEVER PUT SAND DOWN THE SINK) and reshape the sand as before.

11. Record all general observations about stream activity and see if you can identify any stream features such as meanders, braids and so on (refer to text).

DATA and OBSERVATIONS:

Observe very carefully and record these observations in FULL SENTENCES in point form. Sketch any shapes or features of interest.

CONCLUSIONS:

Write a full conclusion with discussion about the observations made during this experiment.

1. What are the main features of rivers seen in this simulation?
2. What happens to the width of the stream's course as it moves downslope?
3. How is the sediment in the stream carried?
4. Describe the motion of the sediment at the (a)start and (b) end of the river
5. Discuss the use of such models (as in this experiment) in the study of real river environments.
6. Why was the sand moistened BEFORE the experiment?
7. What errors could limit the use of such a model? Give some examples from this experiment
8. How such an experiment could be improved.

EXPERIMENT 6.3   2 Lessons

SIMPLE SOIL TESTING EXPERIMENTS

AIM: To perform some basic experiments on soils.

MATERIALS: Soil samples (local and others if possible), Barium Sulfate powder, Universal Indicator in dropper bottles, distilled or de-mineralized water, test tubes (large with stoppers) and test tube racks, clay (whole and powdered), sand, loam, plastic teaspoons, 100 ml measuring cylinders.

BACKGROUND:

The permeability and porosity of soils can be found using experiments as previously described. Other important factors of the soil, such as its organic component and its profile can be found through observation. This activity describes the practical considerations of the acidity (pH) of the soil and its clay content. A good soil should not have too much inert sand (low nutrition and high permeability of water) nor should it have too much clay (low permeability of water). There should also be the right amount of acidity and good topsoil with a high organic component for nutrition.

PROCEDURE:

PART A: Measuring the Acidity (pH) of the soil

1. Place about 1 cm. of soil into a clean test tube.

2. Add about the same amount of Barium Sulfate powder to flocculate (clump together) any clay particles.

3. Add 10 ml of distilled water.

4. Add several drops of Universal Indicator and, placing a thumb over the top of the test tube, shake the mixture.

5. Place the test tube in the rack and allow the solids to settle for a few minutes.

6. Compare the colour of the water above the solids to the colour chart for pH.

7. Repeat for several soils if available.

PART B: Clay and Soil - the Farmers' Technique

Having too much clay clogs up the soil and makes water difficult to penetrate; too little and the soil becomes too loose. A quick test for the texture of the soil by farmers on the land is to:

1. Take a small piece of clay onto the palm of the hand and wet it slightly (a farmer might use spittle but here tap water is preferred).

2. Roll it into a small ball.

EXPERIMENT 6.3 continued

PROCEDURE: continued

3. Use the fingers to knead it out into a long ribbon (some farmers roll it out into a long cylinder with the palm of the other hand).

    (Pure clay should give a long, unbroken ribbon or cylinder)

4. Now repeat this with a sample of soil, noting the maximum length obtained until the ribbon or cylinder just starts to break.

    If your ribbon measures less than 2.5 cm long before breaking, you have **loam** or **silt**.
    If your ribbon measures 2.5 to 5.0 cm long before breaking, you have **clay loam**.

PART C: The Effect of Lime on Clay Soils

1. Place about a teaspoon of powdered clay into each of two large test-tubes.

2. Fill both test-tubes with water to about halfway from the top. Stopper and then shake each to disperse the clay.

3. Add limewater (calcium hydroxide solution) to one of the test-tubes, stopper and shake it. Watch what happens carefully.

4. Also stopper and shake the other test-tube and then place both test-tubes into a rack and allow to stand.

PART D: The Components of the Soil

1. Place about 200 ml of soil into a 100 ml measuring cylinder.

2. Fill it up to the 100ml mark, cover the top tightly with the hand and then shake and invert the cylinder.

3. Stand the cylinder upright and allow the mixture to settle. This may take some time (perhaps overnight).

4. When the components have settled (and the water above is clear), measure each component (sand, silt, clay, organics as dark material) as a percentage of the total height of the solid mass in the bottom of the cylinder e.g. if the dark top layer of the organic component is about 2 ml on the scale, and the total mass takes up 200 ml, then the percentage is 2/200 x 100 % = 1.0%.

5. List the components as a table.

6. Repeat with other soils if available.

EXPERIMENT 6.3 continued

DATA and OBSERVATIONS:

PART A: pH Test

1. What is the pH of the soil?

2. What does this means in terms of acidity?
(Answer for each soil tested - if several soils are used give the data in the form of a table).

PART B: Clay and Soil - the Farmers' Technique

1. Describe the nature of the soil sample used.

2. Is this a good test to use? Why?

PART C: Effects of Lime on the Soil

1. What happens when the lime is added to the clay mixture?

2. What effect (if any) will this have on the dry clay soil after treatment?

3. Calcium hydroxide is alkaline. What will this do to the pH of the clay soil? Would this be harmful to plants grown in this soil (in general plants need a pH of about 5.5 to 7.0)?

RESEARCH: (Optional) How can the pH of a very alkaline soil be reduced?

PART D: Soil Components

1. What are the relative percentages of each soil component?

2. How could this soil sample be classified? (see text book)

CONCLUSIONS:

1. Write a full conclusion with discussion about the observations made during this experiment.

2. What are the main factors (a) shown in this experiment and (b) from additional **research** which are important in understanding about the usefulness of soils?

3. What is pH?

4. What is the original source of (a) clay and (b) the organic matter found in some soils?

5. Are there any improvements or additions which should be made in this experiment?

EXPERIMENT 6.3 continued

CONCLUSIONS: continued

RESEARCH: (Optional)

1. What is the soil type in your local area? Does it need additional improvement to grow crops?

2. What is Hydroponics? How can this technique be used on a small and large scale?

Chapter 7: Landforms

| EXPERIMENT 7.1 | One Lesson |

KARST SIMULATION

AIM: To show a simple model of how limestone caves form below ground.

MATERIALS: Sugar cubes, sand, clear plastic cups, modeling clay, toothpicks and beakers

BACKGROUND:

Limestone caves systems are formed when water containing dissolved carbon dioxide gas (forming a very weak carbonic acid) percolated through topsoil and into the many cracks in the limestone below. These cracks form when the rock is being compressed and consists of horizontal bedding planes and vertical or oblique fissures. Water seeps down these cracks and the weak acid dissolves out the calcium carbonate, the main constituent of limestone. This takes considerable time so another "rock" (sugar cubes) is used instead of the limestone and water (which dissolves the sugar) is used instead of carbonic acid.

PROCEDURE:

1) Add a small amount of sand (say about 2 cm) to a small transparent plastic cup. This sand represents the rock underlying the "limestone".

2) Stack the sugar cubes so that they are staggered and also hard up against one side of the cup (this will be the viewing portal).

3) Fill more sand around the stacked cubes to represent surrounding sediments.

4) Place a thin layer (about 2-3 mm thick) of modelling clay (plasticine) over the top of the cup.

5) Perforate it with the toothpick to give many holes to simulate the porous topsoil.

6) Using a beaker, slowly trickle water over the top of the model and carefully observe (it may be a fast result!). Water can be added until it reaches halfway up the bottom layer of cubes.

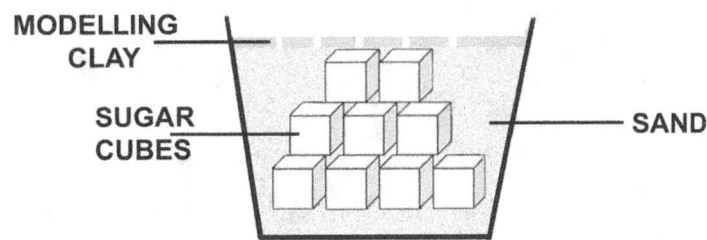

EXPERIMENT 7.1 continued

DATA and OBSERVATIONS:

Observe very carefully and record these observations. SKETCH the model AFTER the simulation has been run.

CONCLUSIONS:

Write a full conclusion with discussion about the observations made during this experiment.

1) Where does most of the reaction take place during the initial stages?
2) Is this consistent with what occurs in nature? Why?
3) What happens at the base of the cubes where water has collected (i.e. below the water table of this model)?
4) What errors could limit the use of such a model?
5) How such an experiment could be improved?

RESEARCH: (Optional)

Use the Internet or other resources to find out about Limestone Caves (a) near your location (b) in other parts where these caves are famous.

1) In some countries such large cave systems are important. Why?
2) What are some of the hazards of living in karst area?
3) What is Spelaeology? Why is it undertaken?

EXPERIMENT 7.2 — One Lesson

THE SHAPE OF RIVERS

AIM: To examine the relationship between the slope of the land surface and the shape of a stream.

MATERIALS: Long white trays, small deep trays, retort stands (rod), boss heads and clamps, 500 ml beakers, sand

BACKGROUND:

The shape and pattern of a river system is determined by the gradient of the land surface which also determines the flow rate. The *Sinuosity Index* of a meandering river is the ratio of the actual river length measured along the entire curve of the stream to the down-valley length or straight line between the two points from which the meanders were measured. This sinuosity index has been used to separate single channel rivers into three general classes: straight (SI < 1.05), sinuous (SI 1.05-1.5), and meandering (SI > 1.5).

PROCEDURE:

1. Set up the following apparatus as in Experiment 6.2:

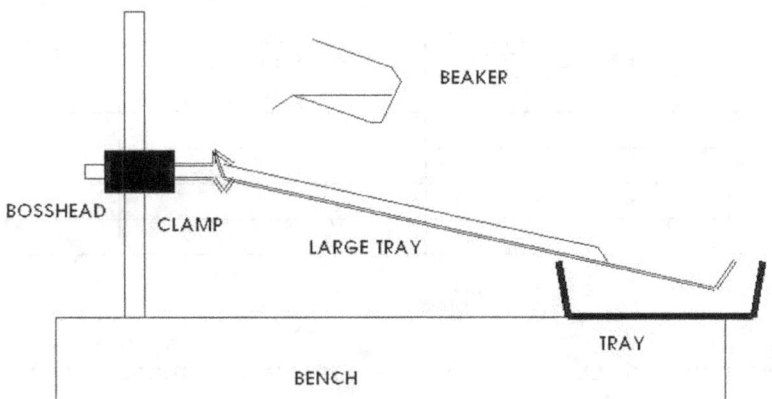

2. Set the large tray to a horizontal (level) position.

3. Add about four beakers full of sand to the tray and moisten until the sand is wet but hard and easy to mold.

4. Smooth the sand out as a flat layer which takes up most of the large tray (as shown above).

5. Adjust the Bosshead so that the tray is at an angle of about 2.5 degrees (Bosshead base is about 21 cm from the bench) so that the flat tray slopes and runs into the smaller tray.

6. Calculate the slope of this surface (change in vertical height/change in length).

7. Using the beaker full of water, gently and consistently pour a continual stream of water onto the sand (as shown in the diagram) from about an height of 10 cm.

8. Make general observations as to: shape of the channel; speed of water; and other general effects on the sand.

EXPERIMENT 7.2 continued

PROCEDURE: continued

9. Measure any meandering sections by measuring the length of the stream (from its centre around its full course) as STREAM LENGTH and also measure the straight line down the stream direction between the two points from which the stream length was measured. This is the COURSE LENGTH.

10. Drain off the water into the collecting tray (or a waste bucket if this is full...NEVER PUT SAND DOWN THE SINK) and reshape the sand as before.

11. Repeat Steps 4 to 9 using different angles e.g. 5.0, 7.5, 10 and 12.5 degrees.

12. Record all general observations about stream activity and GRAPH the sinuosity against angle as a line graph (think about graph construction and which axes to use).

DATA and OBSERVATIONS:

Observe very carefully and record these observations. Sketch any shapes or features of interest. Complete a table of angle and Sinuosity Index and then graph this data.

Angle	Stream Length	Course length	Sinuosity Index
$2.5°$			
$5.0°$			
$7.5°$			
$10°$			
$12.5°$			

CONCLUSIONS:

Write a full conclusion with discussion about the observations made during this experiment.

1. Is there any relationship between the slope of the river bed and its sinuosity? Explain and give reasons.

2. Discuss the use of such models (as in this experiment) in the study of real river environments. Are there other factors which would determine the sinuosity of a river?

3. What errors could limit the use of such a model?

4. How such an experiment could be improved?

EXPERIMENT 7.3 One Lesson

STEREOPAIRS AND LANDFORMS

AIM: To use pairs of stereo photographs (aerial photographs) to interpret features and landforms.

MATERIALS: Pairs of stereo photographs of a variety of landforms and small stereo viewers.

BACKGROUND: Using special stereo cameras (with two lenses at an angle), photographs can be taken from an altitude which will then give a raised 3D image when placed in pairs and view with Stereoscope viewers. The eye sees an impression of raised and sunken features, usually much exaggerated, which often give a better interpretation of a land surface than drawn topographical maps.

PROCEDURE:

1) Use this book or stereo-pairs provided.

2) open out the legs of the stereo viewer and position it on the page or pairs so that the centre of the viewer is approximately over the join of the paired photograph.

3) Whilst looking down through the viewer, move it up and down back and forth across the joining line (of the two photos) until the brain suddenly interprets a raised 3D view (sometimes it helps to open and close alternate eyes to get the view with each eye and then move the viewer until both images overlap).

4) Closely examine the 3D features seen and recognise some of the details of each landform.

PHOTO 1:

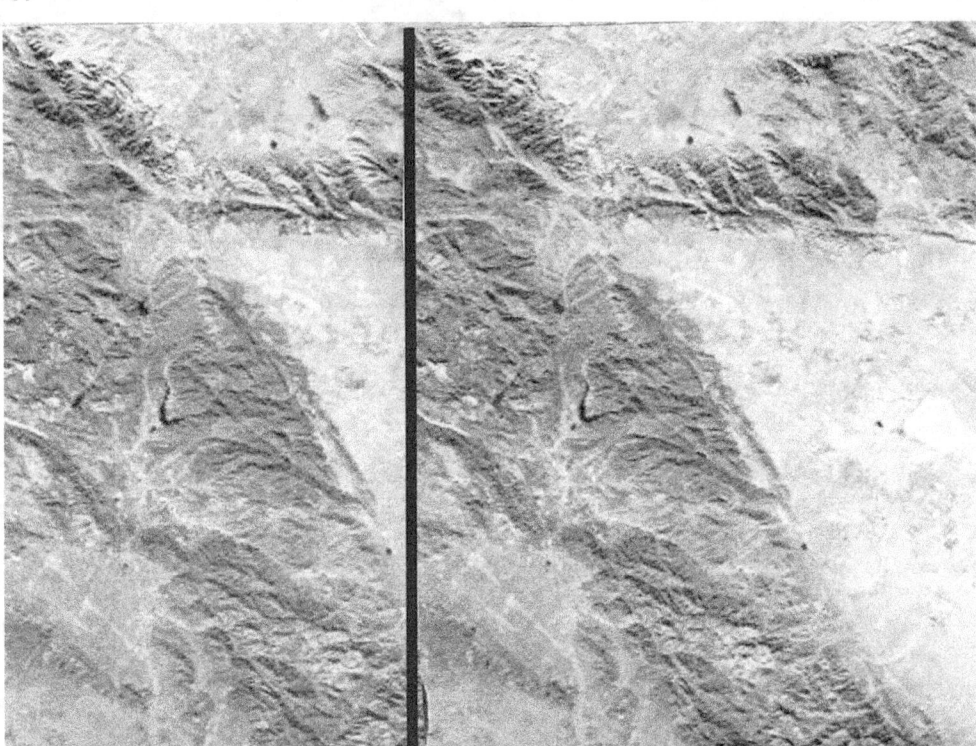

PHOTO 2:

(Photo credit: USGS)

DATA and OBSERVATIONS:

1) What is the main feature or shape seen in each photo?

2) Describe the general land surface of each photo.

3) Suggest a way that these features may have been caused.

CONCLUSIONS:

These photos are low resolution but they can be useful.

1) Comment of the use of satellite and lower altitude aerial photos to locate and describe landforms.

2) What errors would limit the use of stereo pairs?

3) How such a technique be improved?

RESEARCH (Optional): Use the Internet to find out how landforms are studied using more modern techniques such as ANAGLYPHS and 3D Modelling.

Chapter 8: Fossils – Life in the Rocks

EXPERIMENT 8.1 Two lessons

SOME COMMON FOSSILS

AIM: To examine, describe and sketch some common examples of fossils.

MATERIALS: Sets of common fossils (one set/class group), hand lenses or stereo microscopes, pencils and coloured pencils, sheets of paper (photocopying paper is good).

BACKGROUND:

The most basic method of studying fossils is by a thorough examination and sketching of specimens. Considerable care must be put into the examination and sketch (photography is a poor alternative) so that all features which may give clues to the organism's type and environment can be seen. Usually, only a small fragment of the life-form is available and so palaeontologists often have to EXTRAPOLATE (extend data by educated guesses) their ideas using knowledge of modern organisms and their environment. They also use the modern view of life and its surroundings to make guesses about ancient times using the Principle of Uniformitarianism that "the present is the key to the past".

PROCEDURE:

1. Carefully examine each fossil in turn looking for any internal features which can be identified.

2. Each specimen will have an identifying number or label to be written with the sketch and following description.

3. Draw the outline of each fossil in turn onto a clean sheet of paper to a satisfactory scale so that one sketch should take about one-half a page. Some specimens will have to be scaled up to this size by carefully copying the outline shape as near as possible to the original. Use x2, x3 etc. notation to indicate the scale.

4. Use coloured pencils to shade in the specimen by lightly shading using the side of the pencil then smearing it lightly with a finger. A 3D effect can also be given to the fossil by drawing a side around one edge and shading it with heavier pencil.

5. Review the internal structures of the specimen and sketch them within the outline.

6. Using the textbook, the Internet or other resources, identify each specimen in as much detail as possible and give the geological age (Period) to which these organisms belong.

7. Describe each fossil with a short paragraph indicating its SHAPE, INTERNAL FEATURES and if possible the probable environment which would have supported such an organism (e.g. terrestrial, marine, freshwater etc.). They should also be classified as to their type e.g. Plant, Invertebrate (Coral, Echinoderm etc.) or Vertebrate (Reptile, Bird, Amphibian etc.)

EXPERIMENT 8.1 continued

DATA and OBSERVATIONS:

Detailed sketches, descriptions, classification and geological age for each specimen should be given in any order.

CONCLUSIONS:

Write a full conclusion with discussion about the observations made during this experiment.

1. Compare and contrast the use of originals and replicas in this study if appropriate. (Are replicas useful? Why?)

2. Why must the scale be given for each specimen?

3. Give uses for such a study of fossil specimens.

4. What are some problems with such a study? Could it be improved?

5. If this were a major work, what other research activities would be necessary to fully give more precise details of exact species, age and environment?

RESEARCH (Optional): Use the Internet or other resources to

1. Find out about how palaeontologists work in the (a) field and later in (b) the laboratory.

2. Where are the major fossil beds located nearby (with the State).

EXPERIMENT 8.2 One lesson

CORRELATION EXERCISE

AIM: To use fossils to correlate sedimentary beds at different locations.

MATERIALS: Coloured pencils   ruler   glue   sheets of blank paper

BACKGROUND:

Correlation is the comparison of two or more sequences of rock to see if they had been formed at the same time. This can be done by comparing the fossils within them or by looking at the separate rock units (layers or grouped layers) or both within the sequence of rocks at each locality. This can be done on an international scale or at a more local level.

Correlation at a local level is used when geologists map how rocks extend below the ground. To support their findings about surface rocks, geologists will employ drilling teams to drill below the surface. The drills are hollow lengths of pipe which are screwed into the ground by the motor of the drill rig. These lengths of pipe are then brought to the surface, uncoupled from the drill rig and opened to obtain the drill cores which can then be stored in a drill Core Library.

Geologists who are interested in what is below the ground at these localities, must measure the different types of rock in each core, noting the details and any structures or fossils present and constructing a Stratigraphic Column of the data. This is called drill core logging.

PROCEDURE:

1. Carefully examine all of the rock units containing fossils (named as **F1**, **F2** etc.) within each of the columns given on the next page. These columns represent rocks at four different localities which may have had different formations:

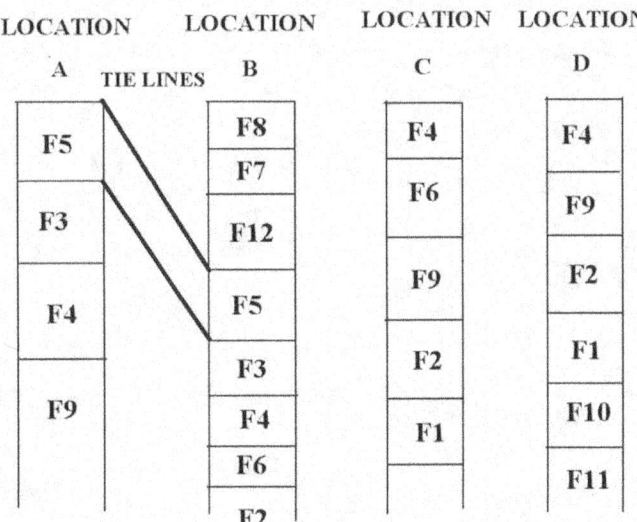

Stratigraphic columns for four localities.
Codes F1-F12 represent fossils and the thickness of the beds which contain them.

EXPERIMENT 8.2 continued

PROCEDURE: continued

2. Copy or photocopy this diagram and then join beds which have the same fossils by using TIE LINES to the tops and bottoms of these beds (one has been done as an example).

3. Looking at the overall connection, determine which are the oldest and youngest beds using the LAW OF SUPERPOSITION (younger beds are placed on top of older beds).

4. LIST all of the beds IN ORDER from **oldest** (which was deposited first) to **youngest**.

DATA and OBSERVATIONS:

Copy, cut and paste the stratigraphic columns and their tie lines to show which beds contain the same fossils and are therefore of the same age. If required scan or photograph this arrangement.

LIST the sequence of the beds from oldest to youngest.

CONCLUSIONS:

1. What ERRORS could occur in using this method to determine relative age of beds?
2. How important is the accuracy of identification of the fossil specimen in this process?
3. Give an hypothesis for why Fossil **F6** is missing from columns A and **D**.
4. This exercise has fossils in all rock units. This is not common as fossils are very difficult to find. Give some reasons why fossils may not be found in rock units at all.

Chapter 9: Economic Minerals and Mining

EXPERIMENT 9.1   Two lessons

SOME COMMON ECONOMIC MINERALS

AIM: To describe, draw and then identify some common economic minerals (Ores).

MATERIALS: Class sets of ore specimens (any of: native copper, galena, sphalerite, pyrite, bauxite, malachite/azurite, haematite, chalcopyrite, pyrolusite, or barites), hand lenses, streak plates, Mohs' scale of hardness kit.

BACKGROUND:

Ores, as minerals can be described using the same PHYSICAL and CHEMICAL PROPERTIES. The most useful properties in identifying specimen include:

1. **COLOUR** - the frequencies of light coming from the specimen when viewed in normal white light. Simple colours (e g black, red-brown etc) are used to describe this property. A mineral may have a variety of colours.

2. **STREAK** - the colour of the powdered mineral. If soft, the specimen may be quickly drawn across a STREAK PLATE (on unglazed tile -white for coloured minerals, black for white or clear minerals). If the specimen is hard, it may be scratched with a knife or a harder mineral and the colour of the scratch observed.

3. **CRYSTAL FORM (Habit)** – the overall appearance of the specimen in the way that it has been formed e.g. CLEAVAGE BLOCK (individual crystals not seen, but many flat crystal surfaces observed giving a blocky appearance); BOTRYOIDAL (many rounded lumps together); FIBROUS (long strands or fibres); RADIATING (long crystals radiating outwards from a common point); GRANULAR (grouped like grains of sand); AMYGDALOIDAL (crystals growing in holes in rock); FOLIATED (as sheets); PISOLITIC (pea-sized spheres); or MASSIVE (no crystals seen).

4. **LUSTRE** -the way light reflects off the specimen's surface. It may vary from surface to surface (i.e. a specimen may have a range of lustre). Lustre may be:

    - METALLIC -hard, shiny like a metal
    - SHINY – very reflective like gloss paint
    - VITREOUS -glassy, with some depth
    - PEARLY -like a pearl or button
    - SILKY -shiny with a fibrous look
    - GREASY -oily appearance like soap
    - RESINOUS -with a dull transparent look
    - DULL -very little shine

5. **CLEAVAGE** -the ability to break naturally along flat surfaces. The best test is to actually cleave the mineral but this is often not desirable so one has to look for the number of cleavage planes which may meet together. Minerals may have:

EXPERIMENT 9.1 continued

BACKGROUND: continued

- NO CLEAVAGE or cleavage in
- ONE DIRECTION (called basal cleavage which gives sheets,
- TWO DIRECTIONS giving lines of small steps,
- THREE DIRECTIONS giving points or corners (which also may be cut off) or even in
- FOUR DIRECTIONS giving four-cornered points like a pyramid (rare).

6. **HARDNESS** - resistance of the mineral to be scratched. Usually compared to a standard set of minerals (MOHS' SCALE) or a set of material for approximations (FIELD SCALE):

MOHS' SCALE	FIELD SCALE
1. TALC (softest)	Thumb nail 2.5
2. GYPSUM	Coin 3.5
3. CALCITE	Glass 5.0 -6.0
4. FLUORITE	Knife 5.5 -6.0
5. APATITE	File 6.5- 7.0
6. ORTHOCLASE	
7. QUARTZ	
8. TOPAZ	
9. CORUNDUM (ruby)	
10. DIAMOND	

7. **SPECIFIC GRAVITY** -the density of the mineral compared to that of water, i.e. how many times the mineral in heavier than an equal volume of water (density of water = 1g/cc). Heft is a word to generally describe the mineral's heaviness in vague terms e.g. heavy, medium, light).

8. **DIAPHEITY** – the way light passes THROUGH the mineral. A most important factor in some gems is that light will pass through them. Diaphaneity can be:

- TRANSPARENT - light passes through unchanged (you can see through it)
- TRANSLUSENT - light goes through but distorted or in part or
- OPAQUE   - light does not pass through at all.

e.g. window glass is transparent, frosted glass is translucent and stone is opaque.

9. **CHEMISTRY** – the general chemical group (e.g. sulfides, carbonates, sulfates etc.) as found by chemical tests (not tested here) and found from reference tables.

EXPERIMENT 9.1 continued

PROCEDURE:

1. Draw up a TABLE (with a full page turned sideways) under RESULTS using the following headings:

   CODE   COLOUR   STREAK   LUSTRE   CLEAVAGE   HARDNESS   OTHER   NAME

2. Test and describe each specimen.

3. DRAW a selection of FOUR specimens to scale (e.g. x2 or x3 etc.) and in colour where appropriate and label main the features seen in each specimen.

4. Use the text, display minerals and data bases to NAME the specimens.

DATA and OBSERVATIONS: (Table and Sketches)

CONCLUSIONS:

1. LIST the code numbers, name and at least two <u>distinctive</u> properties of each mineral which will help in the quick identification of that mineral.

2. Also give the main uses of the ore.

3. Comment on the use of the mineral properties used in this experiment (giving advantages and disadvantages) in identifying unknown ores.

Research (Optional): Use the Internet and other resources to find out:

1. What are the ores mined in the local area or State?

2. How are they mined and where are they processed?

3. What materials other than metallic ores are mined and/or quarried locally? What are they used for?

Chapter 10: Processing the Mined Ore

EXPERIMENT 10.1   **Two lessons**

MAKING COPPER FROM A COPPER ORE

AIM: To extract the metal copper from its ore (malachite) by physical and chemical means.

MATERIALS: small pieces of copper ore (malachite/azurite), geological picks, dilute sulfuric acid, small beaker, measuring cylinder, sandpaper, filter paper with filter funnel and conical flask, clean iron nails, distilled (or filtered) water in wash bottles.

BACKGROUND: Once an ore has been mined and separated from unwanted minerals (gangue) and rock, it must be processed to obtain the metal. This usually is very difficult as the metal in the ore has to be chemically separated from the rest of the elements in the chemical compound of the ore.

Copper can be found in several ores and even as Native Copper. One copper ore is Malachite/Azurite, a mixture of two different crystal forms of the one complex chemical, Copper Carbonate – Copper Hydroxide (Malachite is the green form and Azurite is blue). Both minerals are usually found together and are often deposited from hot water solutions in veins and porous rocks.

Crushing is usually the first step followed by chemical extraction of the metal as a solution. As a carbonate, Malachite/Azurite can be dissolved in acid to give a solution and carbon dioxide gas. If another metal which is more reactive than copper (e.g. Iron) is placed into the copper solution, the copper in solution, will be replaced by the active metal. The copper and will be seen as a coating on the more reactive metal. This coating can be removed as pure copper.

PROCEDURE:

1. Crush the malachite (and the rock which contains it) into fine powder carefully by placing the specimen between sheets of scrap paper on the floor and lightly tap it with the geology pick.

2. Transfer the smaller pieces into a small beaker or flask and add about 50 mL of dilute sulfuric acid (CARE!).

3. Observe and describe any reaction. Wait until all reaction ceases, then add about 50mL of water to dilute any acid which remains.

4. Set up the filtering apparatus with the folded filter paper(fold in half, then half again and open out one of the edges to form a cone of paper) place inside the filter funnel:

EXPERIMENT 10.1 continued

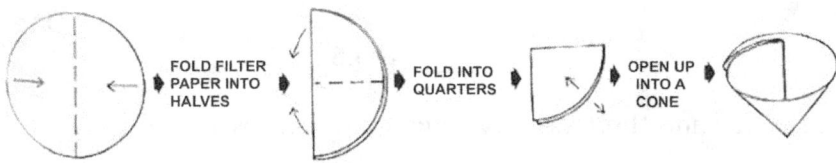

How to fold filter paper

5. Rinse out the filter paper, funnel and conical flask with distilled water which may be poured down the sink.

6. Stir the mixture in the beaker and slowly pour it into the filter paper in the funnel ensuring that the level is never near the top of the filter paper (preventing overflow).

7. The liquid passing into the conical flask should be transparent. If not wait until this step has been completed, add the filtrate (the liquid) into a clean beaker and repeat the last step with new filter paper and clean conical flask.

8. Obtain some iron nails (about 4-5) and ensure that they are shiny by using sandpaper to get a clean surface. Place these into the copper solution which has been filtered and wait.

DATA and OBSERVATIONS:

1. Sketch the filtering apparatus in TWO dimensions (as a standard laboratory diagram).

2. Describe any changes which occur.

CONCLUSIONS:

1. Why was the filter paper rinsed out before the experiment?

2. Write a WORD EQUATION for the reaction between copper carbonate and the sulfuric acid.

3. Write a word equation for the replacement reaction between the copper solution and the iron nails;

4. How could the copper be recovered (removed for use) from the iron nails?

Research (optional):

1. Where in your State or Nation are there (a) copper mines and (b) refineries?

2. LIST four separate uses for the metal copper in society.

Chapter 11: Fuels and Energy

EXPERIMENT 11.1   One or two lessons
FOSSIL FUELS

<u>AIM:</u> To describe, draw and then identify some common fossil fuels.

<u>MATERIALS:</u> **Class sets of** fossil fuels e.g. lignite (brown coal), bituminous coal (black coal), anthracite (hard coal) and oil shale. Hand lenses.

<u>BACKGROUND</u>:

Fossil fuels are sources of energy which have been produced by the chemical breakdown of ancient life-forms. Coal has formed by the breakdown of plants in freshwater with several types of ranks of coal being formed depending upon the completeness of the chemical process and compression of the ancient vegetation.

Oil and natural gas has come from the complex breakdown of marine organisms which fell to the mud of the ancient seafloor. Oil and gas usually leave their source rock and move upwards (due to low density) through porous rock until they are trapped in non-porous structure or come to the surface (tar pits) and evaporate.

Most countries of the world rely on fossil fuels for energy and for raw materials for synthetics, paints and dyes, fertilizers, explosives, plastics and most materials in daily use. Many countries have reserves of fossil fuels to last for hundreds of years, but pollution (mainly of carbon dioxide and monoxide gases) has become a major problem.

<u>PROCEDURE:</u>

1. Use a hand lens to carefully examine each of the specimens for their physical properties such as:

    - COLOUR – describe it as you see it;

    - LUSTRE - how light reflects off the specimen (shiny or dull or glassy)

    - FRIABILITY – is it easy to crumble (friable) or not (non-friable)?
      Do NOT try to break the specimen. Holding it in the hands will show if it will crumble.

    - STRUCTURES – what shapes can be seen in the coal e.g. layers, bands, veins of minerals etc. (use own terms).

    - FEEL - rub a finger over a surface. Is it dry, oily, rough, smooth etc. or any combination of these?

2. Draw up a TABLE in which these descriptions can be given.

3. Make a sketch, to scale (e.g. X2, X3 etc.) and in colour or shading. Label any detail as appropriate.

EXPERIMENT 11.1 continued

DATA and OBSERVATIONS: (Table and Sketches)

CONCLUSIONS:

List each of the fossil fuels and write two or three KEY WORDS to aid later identification.

1. Of the coals, which one has the highest CARBON CONTENT?

2. What other impurities can be in the brown and black coal (see textbook)?

3. How is the oil contained in the OIL SHALE (research needed)?

Research (Optional):

Use the Internet or other resources to:

1. Locate major sources of fossil fuel in the local or State area.

2. Prepare a list of the products OTHER THAN ENERGY obtained from
    (a) coal and (b) oil.

3. Prepare a talk FOR or AGAINST the continued use of fossil fuels

4. Briefly outline what alternatives to the uses of fossil fuels are being used locally.

EXPERIMENT 11.2
ENERGY CONTENT OF FUELS
One or two lessons

AIM: To demonstrate a method of determining the energy content of liquid fuels.

MATERIALS: Small spirit burner, small conical flask, stand and clamp, ethanol alcohol, kerosene, electronic balance, 100 ml measuring cylinder, clock/watch with seconds.

BACKGROUND:

Not all fuels have the same energy content and it may not be efficient to use one fuel instead of another. Pollutants from the fuel and its original source and methods of extraction are other considerations when selecting the type of fuel to use.

PROCEDURE:

Set up the following apparatus:

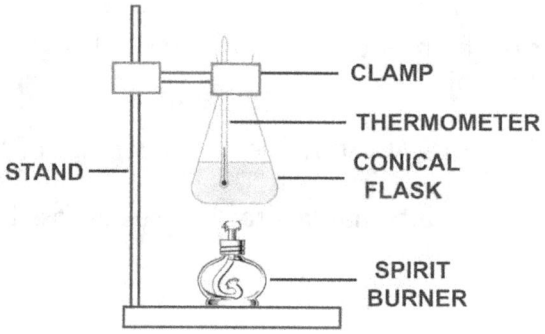

1. Half-fill the spirit lamp with alcohol and weigh it and its cap on the balance and record the total weight.

2. Measure 100 ml of water and place it into the conical flask and measure the temperature of the water with the thermometer.

3. Carefully light the spirit burner and place it under the flask. Start recording time in minutes and seconds.

4. Heat the water so that there is a temperature difference of about 40 – 50 degrees (but do NOT boil the water). Measure the water temperature and the time taken.

5. Extinguish the flame of the spirit burner with its metal cap and when cool, re-weigh the burner and its cap to determine how much alcohol fuel was used.

6. Make sure that there are no naked flames in the room.

7. Remove the alcohol into a waste container and half fill the spirit burner with kerosene (paraffin or lamp oil)

8. Repeat steps 1 to 5.

9. Calculate the heat required to raise the temperature of the water using the equation (below) and therefore the amount of heat per kilogram of each fuel.

EXPERIMENT 11.2 continued

DATA and OBSERVATIONS:

Re-draw the diagram as shown above.

Complete a table for alcohol and kerosene (paraffin) showing:

MASS of WATER (kg), INITIAL TEMPERATURE (C^0), FINAL TEMPERATURE (C^0) and CHANGE in TEMPERATURE ΔT (^0C) and TIME TAKEN

CALCULATIONS:

There are two parts to the calculation and a third to do with Power:

- Finding the energy given to the water (as heat) in Joules by calculating it using the formula:

    HEAT = MASS x SPECIFIC HEAT CONTENT of WATER x TEMPERATURE CHANGE

    or in symbols: $Q = mc\Delta T$

    where Q = heat content (Joules - J)
    m = mass (kilograms)
    c = specific heat of the substance (J/kg.^{0}C) and
    ΔT = change in temperature (^0C)

    For water, c = 4,184 Joules/kilogram ^0C

    Also, 100 ml. of water weighs 0.1 kilograms.

- This only gives the amount of heat theoretically given to the water by burning the fuel, so now divide this heat value (converted to mega joules by dividing by 1000) by the mass of fuel used (in kilograms).

    Heat Content MJ/kg) = $\dfrac{\text{Amount of Heat Energy (MJ)}}{\text{Mass lost by fuel (in kg)}}$

- Also, the POWER (P) in watts (W) of this system is the amount of ENERGY (heat) in Joules (J) divided by the TIME (in seconds) at which it was applied:

    Power (W) = Energy (J)/Time (s)

EXPERIMENT 11.2 continued

CONCLUSIONS:

1. What was the calculated value for the Heat Content (or Specific Energy) of:

    (a) ethanol and (b) kerosene(paraffin)?

2. How do these experimental values compare to the real values in scientific literature (23.4 to 26.8 MJ/kg for ethanol and 46.2 MJ/kg for kerosene)? What errors are there in this experiment?

3. What was the Power of the system for both fuels?

4. How could that Power be increased?

5. How could this experiment be improved?

Research (Optional):

1. Use the Internet to find out about James Watt and James Prescott Joule.

2. Often people talk about calories or the calorific value of foods. What is a calorie and how does it relate to the kilojoule (the more modern unit also given for foods).

3. Research how these heat measurements are made in the laboratory (Hint: look up calorimeter experiments)

Chapter 12: Exploring the Seas

EXPERIMENT 12.1 One lesson and non-class time

INTRODUCTION TO NAVIGATION

AIM: To introduce some simple concepts of navigation.

MATERIALS: Simple DIY inclinometer (large protractor, length of wood or cardboard, plumb bob) or retort stand and clamp, ruler, protractor, string and blu-tac (sticky clay).

BACKGROUND:

Navigation is both a science and an art. Early seafarers could find their LATITUDE (angular distance from the Equator) and NORTH using the stars and the Sun. At night, they would mostly use POLARIS (the North Star – see p. 406 of text) in the Northern Hemisphere and much later, the SOUTH CELESTRIAL POLE (using the Southern Cross constellation, p.406) to find the angle from these points in the sky to the Horizon. Using the Sun to find the Latitude in the day was more difficult because it had to be done when the Sun was at its highest point in the sky (12 Noon). Seafarers used various instruments: the Jacob's Staff, the Astrolabe and later the Quadrant and the Sextant to measure the angle between the noonday Sun and the horizon.

Now the Sun also changes its position north and south throughout the year, swinging from being overhead at the Tropic of Cancer (Northern Hemisphere Summer Solstice/Southern Hemisphere Winter Solstice – about June 20th to 22nd), across the Equator (the Equinox occurring twice about 20 March and 22-23 September when it swings back again) to the Tropic of Capricorn (Summer Solstice in Southern Hemisphere/Winter Solstice Northern Hemisphere between December 20 and December 23). This means that any noon sighting of the angle of the Sun to the horizon must also be compensated or adjusted for the time of the year. To complicate matters even further, the Earth also wobbles slightly (precession) so these angles also change slightly each year. To make these adjustments, seafarers would consult a book of tables (a Nautical Almanac) for that particular year and date. These tables were devised by astronomers who had measured these Sun (and star) changes over centuries and as they formed a recurring set of patterns, could make predictions about future years and publish Almanacs in advance.

Longitude is the angular distance around the globe East to West. This was a difficult problem and was not solved until late in the 18th Century by the invention of Harrison's nautical chronometer (a very accurate, sea-going clock). Before then, seafarers would stay in sight of land or use their compass (or the stars) to determine their course direction (bearing) and a set number of days to sail along this bearing (from past experience). This is called DEAD RECKONING and often they would be ship wrecked if currents or winds took them too far or off course during this time. Today, longitude is measured EAST or WEST from the Prime Meridian (0^0) at Greenwich near London, United Kingdom.

Now, using the method below to get an angle for the Sun's altitude would only be valid if the Sun was at the equator (at an Equinox) on the dates about 20th March and 20th September. The tilt of the Earth between 23.4 N (the Solstices about 20th June in the Northern Hemisphere and December in the Southern Hemisphere) so one must calculate the difference in angle due to the Earth's tilt of 23.4 degrees by adding or subtracting the angle for today's date. The table (below) gives the approximate corrections to be made for dates between one Solstice and the next.

EXPERIMENT 12.1 continued

DATE	+23.5°	+20	+15	+10	+5	0	-5	-10	-15	-20	-23.5
DECLINATION	June	May 21 & July 24	May 1 & Aug 12	April 16 & Aug 28	April 3 & Sept 10	March 21 & Sept 23	March 8 & Oct 6	Feb 23 & Oct 20	Feb 9 & Nov 3	Jan 21 & Nov 22	Dec 22
	Solstice					Equinox					Solstice

Remember that the Summer Solstice is when the Sun is directly overhead at the Tropic of Cancer (Northern Hemisphere) or the Tropic of Capricorn (Southern Hemisphere) and the Equinoxes (occur twice) are when the Sun is overhead at the Equator. For example, if this experiment was being performed on the 3rd of April, then the declination would be + 5 degrees.

Today, ships and small-boat sailors use GPS systems to find their locations very accurately but any good mariner will still carry a sextant and current Nautical Almanac and know how to use them.

PROCEDURE:

PART A: LATITUDE

1. Make a simple Astrolabe inclinometer using a retort stand, clamp, ruler, protractor and a plumb bob made from string and a piece of clay or some other small weight:

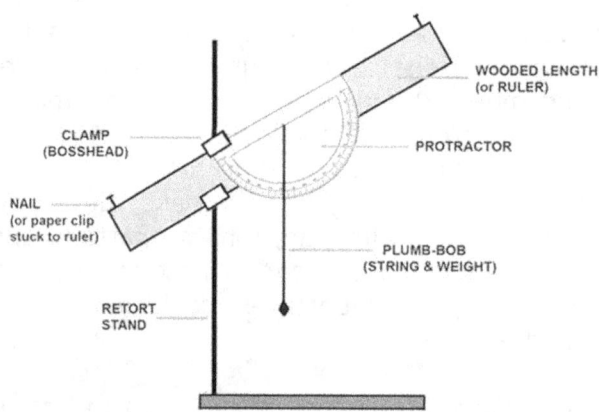

2. **At 12 noon,** or as close to it as possible, place the inclinometer on a flat surface in direct sunlight and rotate the base AND wooden length so that it faces the Sun. It will be exactly on target when the shadows of the two nails exactly overlap each other.
   DO NOT LOOK DIRECTLY AT THE SUN

3. Read off the angle shown on the protractor (it will be the smallest angle) and record the value. This is the elevation angle (e).

4. Return to the classroom and obtain the class average for this angle.

EXPERIMENT 12.1 continued

PROCEDURE continued:

PART B: LONGITUDE:

Longitude is found by the time difference between the observer's local time and that at Greenwich. Sailors would carry a very accurate sea-going clock (called a Chronometer) which had been set at Greenwich Mean Time. Sometimes this would mean finding local noon (12 noon) by noting when the Sun was at its highest peak above the ship (found by several sightings over time using a sextant or shadow stick)

1. Find out the local time by carefully observing a clock in the classroom. Perhaps the most common time can be taken from a number of clocks, watches or cell phones/computers in the room.

2. Use the Internet to find out Greenwich Mean Time (search for Greenwich Mean Time now).

3. Note whether or not this time is ahead of the Greenwich Mean Time (i.e. EAST of it) or past it (WEST).

DATA and OBSERVATIONS:

Draw a sketch of the diagram for sighting the Sun.

PART A: LATITUDE:

1. What is the measured angle of the Sun?

2. What is the group average?

3. What is the declination of the Sun for today's date:

4. Which Hemisphere were the observations made (i.e. NORTHERN – USA, Europe, UK etc. or SOUTHERN – Australia, South Africa, New Zealand)?

PART B LONGITUDE:

1. What is the local time (averaged):

2. What is Greenwich Mean Time:

3. What is the time difference between the two?

EXPERIMENT 12.1 continued

CALCULATIONS:

PART A: LATITUDE:

For Northern Hemisphere students:

$$\text{Latitude} = 90° - (e + d)$$

Where e is the angle of elevation taken of the Sun (class average) and d is the declination of the Sun for that date. Remember, if the declination is a NEGATIVE value, this is still added (e + -d).

For Southern Hemisphere students:

$$\text{Latitude} = 90° - (e - d)$$

Remember, if the declination is a NEGATIVE value, this is still subtracted (e - -d) or (e +d)

PART B: LONGITUDE:
For both hemispheres:

Calculate the number of degrees for each hour and fractions of the hour difference from Greenwich Mean Time (remember that each hour = $15°$ of Longitude since the Earth turns $360°$ in 24 hours and 360/24 = 15).

CONCLUSIONS:

1. What is the Latitude (degrees North of South of the Equator 0-90°) and Longitude (East or West of Greenwich 0-180°) of the observer.

2. Discuss the errors which would occur in the measurements in
    (a) Part A and in
    (b) Part B

3. What would be some of the problems when determining latitude at sea (even with a good sextant)?

4. Clocks had been invented well before the 1740's so why was Harrison's Chronometer so important?

5. Today most navigators use a GPS system. How could such a system go wrong?

Research (Optional):

1. Find out about the life and research of John Harrison.

2. What other method of navigation could a ship at sea use?

EXPERIMENT 12.2
MAPPING THE DEPTHS

One lesson

AIM: To use depth recording data to construct a typical profile across an ocean.

MATERIALS: Data provided, graph paper or plain paper ruled, pencils

BACKGROUND:

The use of a technique called sonar (SOund Navigation And Ranging) greatly improved the ability of ships to measure the depths of the ocean (Bathymetry). In SONAR a ship sends out a pulse of sound (a ping), which is reflected by the sea floor and detected with the ship's instruments. If the time it takes for a reflected ping to be heard is measured and the speed of sound in water is known, then the water depth can be derived. Using sonar, ships could record continuous depth measurements without stopping.

In this exercise, an oceanographic ship sails eastward from Puerto Rodrigo to Orupendo Bay, taking SONAR readings underway at various intervals at stations marked on the map. North is at the top of the page.

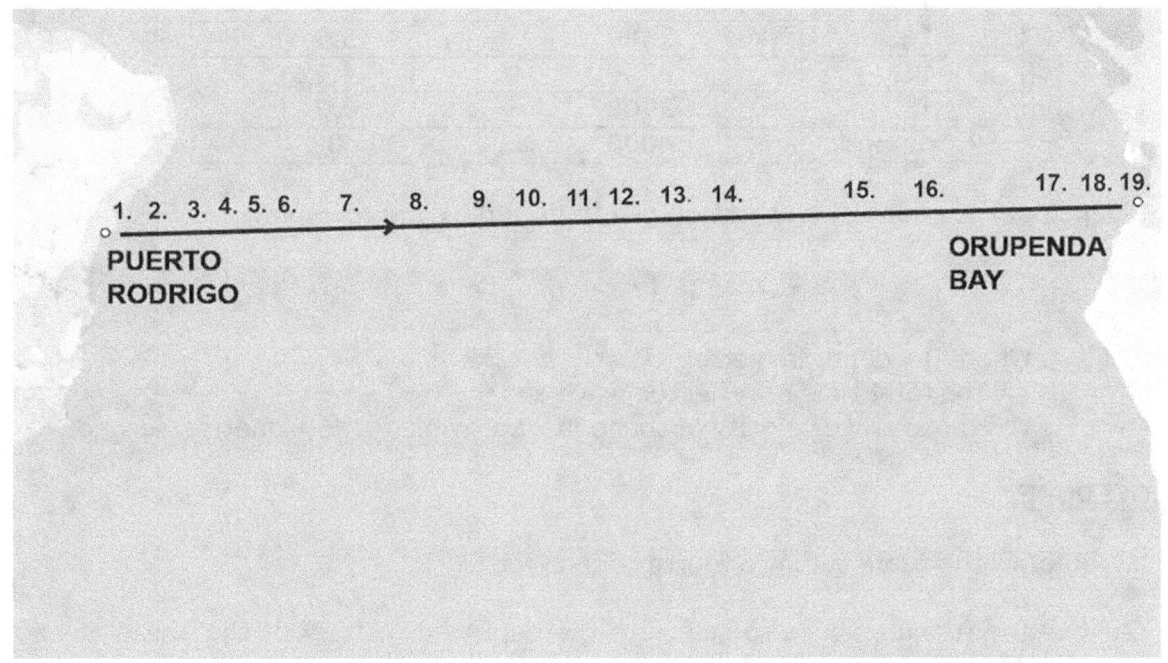

The total time for a reflection to return to the ship was recorded at each station and is given in a table on the next page:

EXPERIMENT 12.2 continued

STATION	DISTANCE TRAVELLED (km)	REFLECTION TIME (seconds)
1. Puerto R.	0	0
2.	100	0.3
3.	200	4.0
4.	280	5.3
5.	400	6.4
6.	600	6.9
7.	1000	6.6
8.	1400	6.3
9.	1800	5.8
10.	2000	5.1
11.	2100	3.5
12.	2200	3.3
13.	2400	5.8
14.	2600	6.6
15.	3200	6.9
16.	3600	5.9
17.	3800	4.2
18.	3900	0.2
19. Orupendo	4000	0

The depth of the ocean at any station can be found by the formula:

$$D = V \times T/2$$

Where D = depth in metres,
T = the total time of reflected wave and
V is the average velocity of sound in water which is 1507 metres/second.

PROCEDURE:

1. Redraw this table adding a fourth column for DEPTH.

2. Use the formula above to calculate the depth (in metres) of the ocean at each station.

3. On a sheet of A4 graph paper or suitable plain paper ruled up, draw a large graph with DEPTH in metres along the vertical axis and the DISTANCE TRAVELLED in kilometres along the horizontal axis.

4. Mark in each station and join the plotted points as a curve of best fit and show sea level along the top of the graph.

EXPERIMENT 12.2 continued

DATA and OBSERVATIONS:

Redraw the table adding the fourth column for depth.
Draw the graph using an appropriate (separate) scale for DEPTH and DISTANCE TRAVELLED. This graph should be drawn with the A4 paper turned on its side and will give a shape which will have a very large vertical exaggeration and thus greatly exaggerated slopes.

CONCLUSIONS:

1. What is the main feature seen in this cross-section of the ocean floor? How was it formed?

2. What other oceanographic methods could be used to confirm the nature of this feature?

3. What name is given for the section of this cross-section:

    a. just offshore east from Puerto Rodrigo and just offshore west of Orupenda Bay?
    b. between Stations 2 to 3?
    c. between Stations 4 to 6?
    d. between Stations 6 to 8 and 14 to 15?

4. What is the slope of the section between:

    a. Stations 2 to 3? (Hint: Slope = Rise/Run and use trigonometry)? And
    b. Stations 4 to 6?

5. What is the general depth of this section in 2 (c) above?

Chapter 13: The Dynamic Earth

EXPERIMENT 13.1   One lesson

HOOKE'S LAW

AIM: To verify Hooke's Law that within the elastic limit of a material (such as springs, rock etc.), stress (applied force/area) is proportional to strain (distortion increase)

MATERIALS: Retort stand and clamp, ruler graduated in millimetres, helical spring, slotted masses and carrier, small bench (or G) clamps.

BACKGROUND:

Hooke's law is a principle of physics that states that the stress (force/unit area) needed to extend or compress a material such as a spring or rock by some distance X is linearly proportional to the strain (or increase in deformation). This will only occur within the elastic limit of the material, that is that limit when it will return to its original length or shape when the applied force has been removed.

In this experiment, the coiled spring represents the material to be stressed by adding weights to its end. The strain placed upon this spring can be measured in terms of the length that the spring extends as the weights are added. So for the spring, Force applied (F) = kX, where k is a constant factor characteristic of the spring (the Spring Constant or its stiffness), and X is small compared to the total possible deformation of the spring.

This law is named after 17th-century British physicist) Robert Hooke (1635 – 1703) and it has been found also to be useful in rock mechanics when discussing the forces which may be applied to rocks causing them distort.

PROCEDURE:

Set up the apparatus as shown below:

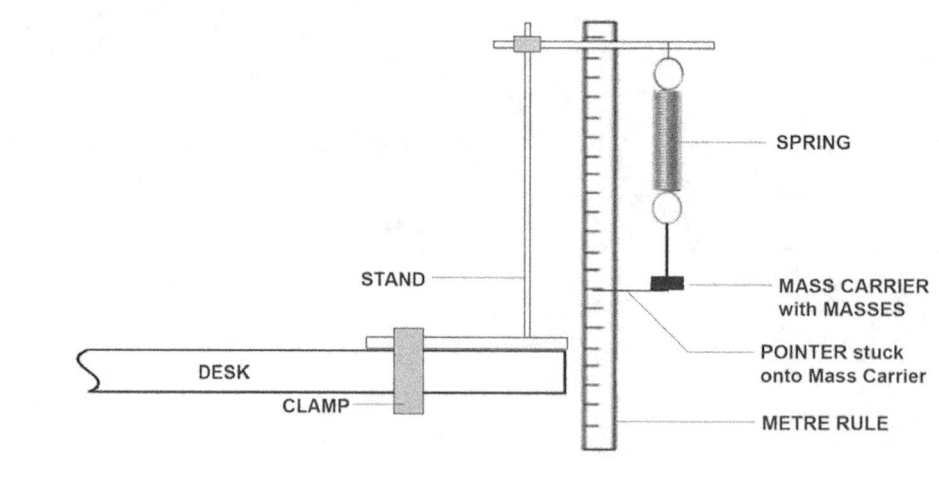

EXPERIMENT 13.1 continued

PROCEDURE: continued

Record the measurement on the metre rule with just the empty mass carrier and its mass (usually 50 grams).

Add successive values of 50 g (0.05 kg) masses to the mass carrier and record the new measurement on the metre rule for the extended length of the spring in a table.

Repeat this addition of mass up to about 400 g (i.e. 0.400 kg) total, measuring the length from the starting position on the metre rule each time.

Calculate the force applied by the addition of the weights by multiplying the value in <u>kilograms</u> by 9.8 (m/sec^2) to give the force units in newtons (N).

Draw up a table showing the values for mass (g), the value for the force applied (kg) and the length of spring from the starting position (in mm).

Graph the values for the force applied on the horizontal axis and those for the length of the spring on the vertical axis.

DATA and OBSERVATIONS:

Redraw the diagram of the apparatus, construct the table as suggested and draw the graph accurately to a suitable size.

CONCLUSIONS:

1. What is the shape of the graph?

2. What does this shape suggest about the relationship between the length of the spring and the force applied to the spring?

3. Is this consistent with Hooke's Law? Explain.

4. What is the value of the Spring Constant (k)?

Research (Optional)

1. Find out about the life and work of Robert Hooke.

2. Why would knowledge of Hooke's Law be useful in geology?

EXPERIMENT 13.2   One lesson or demonstration

DEFORMATION AND TEMPERATURE

AIM: To show the effects of temperature on deformation of a plastic material.

MATERIALS: Retort stand and clamp, ruler graduated in millimetres, Plasticine, slotted masses and carrier, small bench (or G) clamps, thermometers (say 0-100^0), clocks or watches with second hands.

BACKGROUND:

Some materials are plastic i.e. they do not obey Hooke's Law; when they are deformed even slightly, they stay deformed and if stressed too much or suddenly may break.

PROCEDURE:

Note: this is the same apparatus as the previous experiment on Hooke's Law except that the elastic metal spring has been replaced by soft, Plasticine.

1. Set up the apparatus as shown below with a long length of rolled Plasticine fastened top and bottom by clear tape:

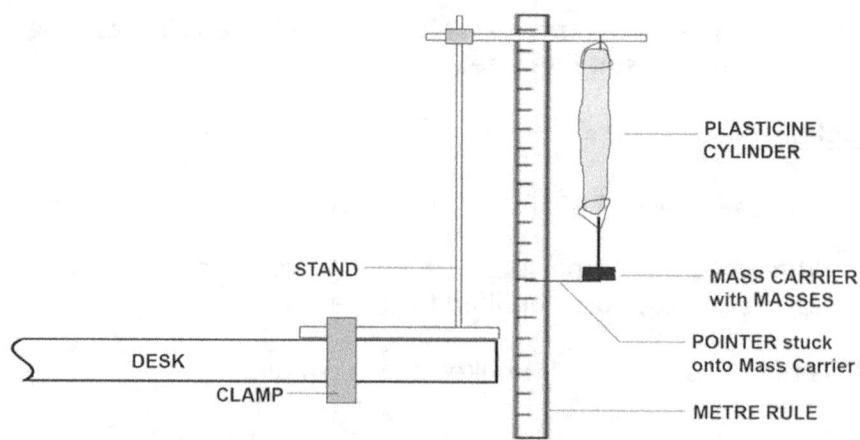

2. Take equal amounts of Plasticine so that four equal cylinders can be rolled to about 6-8 cm in length and about 1 to 2 cm in cross-section.

3. Place a cylinder into a small plastic bag and emerge it into ice water (so that the bag keeps it is dry) for two minutes.

4. Measure the temperature of the water.

5. Quickly remove it from the water and bag and, with minimal handling, attach it to the stand with tape. Attach the mass carrier to its base also with tape.

6. Note the length of the cylinder with the mass carrier attached using the pointer.

EXPERIMENT 13.2 continued

PROCEDURE: continued

7. Add successive values of 50 g (0.05 kg) masses to the mass carrier, timing the experiment until the cylinder breaks. Record the minimal mass required to do this and the length of the cylinder. Also describe the deformation which occurred just before it breaks.

8. Repeat steps 3 to 7 using a cylinder which has been placed in ordinary water from the tap (faucet).

9. Repeat steps 3 to 7 for a cylinder placed in half tap water and half very warm water, being sure to measure the temperature of the water.

10. Repeat steps 3 to 7 using very warm water, recording its temperature.

DATA and OBSERVATIONS:

Redraw the diagram of the apparatus.

Draw up a table showing the: temperature of the water; the length at the start (50g mass carrier); total mass added for each step; the corresponding length of the cylinder; and the time and mass required to break the cylinder.

Describe any deformation fully for each temperature.

CONCLUSIONS:

1. Why isn't Plasticine considered elastic and obey Hooke's Law?

2. Why was the Plasticine placed into bags before emersion in water?

3. Why is minimal handling required when attaching the Plasticine to the stand and mass carrier?

4. What is the relationship between temperature and the mass require to break the cylinder?

5. What is the relationship between the temperature and the time taken to break the cylinder?

6. Describe <u>how</u> the cylinder broke for each of the temperatures.

Research (Optional)

Other than external force and temperature, what other factors may be involved in the deformation of rock such as sedimentary rocks?

EXPERIMENT 13.3   **One lesson or demonstration**

COMPRESSIONAL STRUCTURES

AIM: To show the effects of compression in the deformation of rock layers.

MATERIALS: Small, transparent rectangular plastic boxes, flour (or corn starch), fine sand, wooden rectangular block.

BACKGROUND:

When rock layers are compressed by large, regional Earth forces, they undergo deformation producing a range of structures such as anticlines, synclines, overfolds and sometimes shear faults.

PROCEDURE:

1. Place the wooden block vertically into the rectangular box at one of its ends so that it will act as a piston

2. Spread a layer of fine sand on the bottom of the box and smooth it out so that the top is horizontal.

3. Cover this with a layer of fine flour so that it makes a thin, horizontal layer over the sand.

4. Add another layer of fine sand on top of this and smooth to horizontal.

5. Add another thin, horizontal layer of flour on top of this sand layer.

6. Add the last layer of sand on top. The final apparatus should look like this:

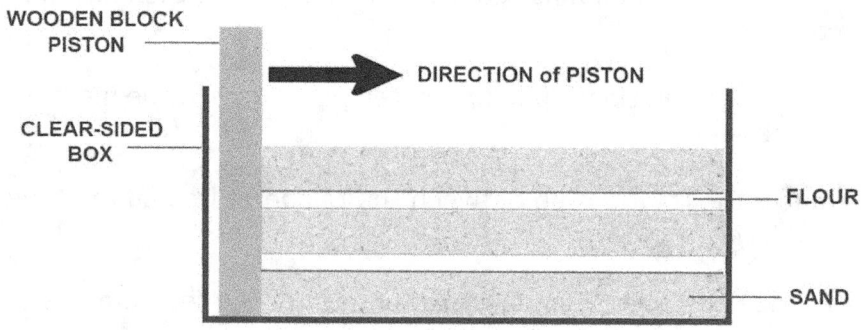

SKETCH 1

7. Carefully move the wooden piston sideways to compress the sand-flour layer slightly (say about one third the length) until slight symmetrical folds are formed. Sketch these folds (SKETCH 2) and give them appropriate names.

77

EXPERIMENT 13.3 continued

PROCEDURE: continued

8. Continue moving the piston until extreme folding occurs and the limbs (sides) of the folds are over each other. Sketch these folds and give then an appropriate name (SKETCH 3).

9. On SKETCH 3, use a ruler and pencil to mark possible FAULT LINES.

DATA and OBSERVATIONS:

Draw SKETCHES 1, 2 and 3, labeling the types of folds and any faults seen in the model.

CONCLUSIONS:

1. Comment on the use of this sandbox model in showing folding and faulting in rock.

2. Relate the extent of the compression to the types of folding produced.

3. How are such models useful in the understanding of a dynamic Earth?

Research (Optional)

Use the Internet to find photographs of the types of folding and faulting seen in the model.

Chapter 14: Earthquakes

EXPERIMENT 14.1	One or two lessons

INTRODUCTION TO SEISMOLOGY

AIM: To locate epicentres of earthquakes using triangulation from seismic waves.

MATERIALS: Computer or tablet with Internet   ruler with millimeters scale
calculator   paper

BACKGROUND: The location of an earthquake is usually given as its EPICENTRE, the place on the surface directly above the source of the earthquake (its FOCUS). Epicentres are found by triangulation from at least three seismic stations using sensitive seismometers and the pattern which they produce called seismograms. Of the three types of earthquake waves (P, S and L), P and S waves travel through the Earth, with the S waves being slower than the P waves. The P waves will arrive first followed by the S waves depending upon how far away the epicentre is from the seismometer detecting the waves. Knowing the speed of each wave, the distance of the epicentre from the seismometer can be calculated. Triangulating at least three values from three widely separated seismometers, the location of the epicentre can be found.

PROCEDURE:

1. Go to the website at:

   http://www.sciencecourseware.com/VirtualEarthquake/VQuakeExecute.html

   This can be accessed by control-left click or by typing this URL directly into the menu bar in a search engine.

   Use the older website as it appears on the screen.

2. When in the Virtual Earthquake site, scroll down to the bottom and click the box for SAN FRANCISO site and then click the button for SUBMIT CHOICE.

3. Follow all of the instructions given in the site and make measurements of the differences between the P and S waves as shown on the screen. Record the three values for each of the seismograms shown in the boxes provided on the next screen as accurately as possible (use an external ruler over the screen to estimate distance between the grid squares on the graphs provided).

4. Go to the next screen and convert these time differences to distances using the Travel Time Graph at left as accurately as possible. Hit the FIND EPICENTRE button to see the triangulation (where the three circles overlap) and get an estimation of the accuracy of these measurements (hit the button VIEW TRUE EPICENTRE).

5. If this accuracy is not good enough it may be repeated for San Francisco by going to REMEASURE P-S WAVES.

EXPERIMENT 14.1 continued

PROCEDURE: continued

6. On completion of the epicentre exercise, scroll down to the bottom of the page and follow instructions for measuring the RICHTER SCALE value for that earthquake.

7. If time permits, go back to the home page and complete a new exercise for another city given in the program.

DATA and OBSERVATIONS:

Record the location of each epicentre found in this exercise and give the Richter Scale for each earthquake.

CONCLUSIONS:

1. Comment on the need for accuracy in making measurements from seismograms.

2. How could this accuracy be improved?

3. How is the energy of the earthquake related to the amplitude of the waves and the distance from the epicentre?

Research (Optional)

Use the Internet to find photographs of the types seismographs and explain how smaller models can be used.

EXPERIMENT 14.2 One lesson

LOCATIONS OF SOME MAJOR EARTHQUAKES

AIM: To plot the location of the epicentres for some major earthquakes to see if there is a pattern in their location.

MATERIALS: World Map   Table of major earthquakes and their locations

BACKGROUND: There are several reasons why earthquakes occur. This activity is an exercise in plotting the latitude and longitude of some of the world's major earthquakes to see if there is a pattern in earthquake locations.

PROCEDURE:

On a photocopied map of the world, carefully plot the latitudes and longitudes for each earthquake given in the table below by using the numbers for these locations. Remember to plot latitude first and then the longitude. Some locations are regions and so may be shaded for the entire length between the positions given:

NUMBER	PLACE or REGION	LATITUDE	LONGITUDE
1	Aleutian Islands	from 165E to 170W	55N 52N
2	Anchorage, Alaska	152W	60N
3	Andes Mountains	from 75W to 70W	5N 55S
4	East Africa	from 40E to 35E	10N 20S
5	Himalaya Mountains	from 70E to 90E	30N 28N
6	Indonesia	from 95E to 130E	5N 10S
7	Kobe, Japan	135E	32N
8	Kurile Islands	150E	45N
9	Lisbon, Portugal	10W	40N
10	Mexico City, Mexico	105W	30N
11	Mid-Atlantic Ocean	0 40W 20W 0 60E 70E	80N 40N 0 50S 50S 0
12	Napier, New Zealand	178E	40S
13	Naples, Italy	15E	40N
14	Papua Nuigini	From 132E to 160E	0 20N
15	Phillipines	from 125E to 120E	5N 20N
16	Salaparuta, Sicily	12E	38N
17	San Francisco, USA	123W	35N
18	Taiwan	122E	22N
19	Tokyo, Japan	140E	35N
20	Turkey	30E	53N
21	Vanuatu	168E	15S

EXPERIMENT 14.2 continued

DATA and OBSERVATIONS:

Plot the location of each epicentre on the photocopy of the map (provided) using the numbers of each location. Students doing electronic reports will have to photograph or scan their completed map.

CONCLUSIONS:

1. Is there any pattern or logical groupings of epicentres?

2. Why do these epicentres exist in these locations?

3. Account for the epicentres in other locations where there might not be a grouping?

4. How could a more accurate world pattern be obtained, especially in locations of isolated earthquakes?

Research (Optional)

Use the Internet to find out about some of the most famous earthquakes e.g.
- Lisbon, Portugal in 1755
- New Madrid, USA in 1812
- San Francisco, USA in 1906
- Southern Chile in 1960
- Tanshan, China in 1976
- Loma Prieta , USA in 1989
- Kobe, Japan in 1995
- Gujurat, India in 2001
- Boumerdès, Algeria in 2003
- Haiti 2010
- Pueblo, Mexico in 2017
- Papua New Guinea in 2018

Chapter 15: Volcanoes

EXPERIMENT 15.1   One lesson

SHAPE OF A VOLCANO

AIM: To draw a topographical cross-section of a volcano and deduce its possible eruptive type.

MATERIALS: Photocopies of the topographical map (below) graph paper, ruler and pencils

BACKGROUND: Volcanoes often have distinctive shapes due to their eruptive type and the material produced by these eruptions, lava, ash or a combination of these.

PROCEDURE:

Look at the topographic map below which shows contour lines for a typical volcano of a particular type. Heights above sea level are given in metres and there is a deep crater lake at the summit.

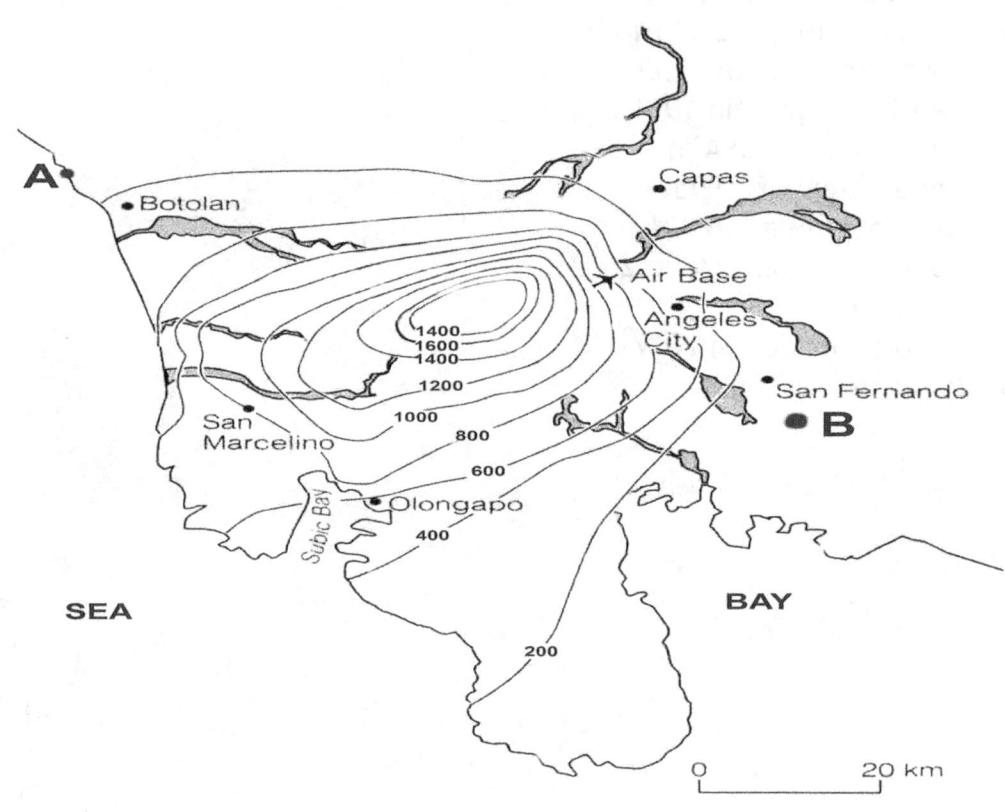

83

EXPERIMENT 15.1 continued

PROCEDURE: continued

2. Draw a topographical cross-section between A and B on the map, choosing an appropriate scale for the vertical height. Students submitting an electronic report should do the cross-section on paper and then photograph or scan it.

DATA and OBSERVATIONS:

Drawn topographical cross-section.

CONCLUSIONS:

1. How is this volcano most likely to be classified (as far as eruption material is concerned)?

2. Give reasons for your classification.

3. If the volcano has not erupted in many years, what would any new eruption likely to be in the first few hours?

4. What would be the main hazards to the nearby cities? Why?

Research (Optional)

1. Use the Internet to find out how volcanoes (especially those near centres of population) are monitored and predictions made about future eruptions.

2. What emergency action would be taken by the local population if there was an imminent threat of this volcano erupting.

EXPERIMENT 15.2

One lesson

LOCATIONS OF SOME MAJOR VOLCANOES

AIM: To plot the locations of some of the world's major volcanoes to see if there is a pattern in their locations.

MATERIALS: World Map   Table of major volcanoes and their locations

BACKGROUND: There are more than 1500 active volcanoes of several different types in the world today. There are many more which have not erupted in living history and so are considered extinct. Occasionally, new volcanoes occur below the sea and on land. This activity is an exercise in plotting the latitude and longitude of some of the world's major volcanoes to see if there is a pattern in the locations of volcanoes.

PROCEDURE:

On a photocopied map of the world, carefully plot the latitudes and longitudes for each volcano given in the table below by using the numbers for these locations. Remember to plot latitude first and then the longitude. Some locations are regions and so may be shaded for the entire length between the positions given:

NUMBER	NAME	PLACE or REGION	LATITUDE	LONGITUDE
1	Agung	Bali, Indonesia	8S	115E
2	Asama	Japan	36N	140E
3	Bezymienny	Kamchatka, Russia	55N	160E
4	Big Ben	Heard island, Australian Antarctic Territory	55S	75E
5	Cotopaxi	Ecuador	1S	78W
6	Misti	Peru	16S	71W
7	Erebus	Antarctica	78S	165W
8	Etna	Sicily, Italy	38N	15E
9	Fuego	Guatemala	15N	90W
10	Fujiyama	Japan	35N	139E
11	Galunggung	Java, Indonesia	5S	105E
12	Heimaey	Iceland	63N	20W
13	Izalco	El Salvador	13N	88W
14	Katmai	Alaska, USA	60N	155W
15	Kilauea	Hawaii, USA	20N	155W
16	Kilimanjaro	Tanzania, Africa	3S	37E
17	Krakatau	Indonesia	7S	105E
18	Lamington	Papua new Guinea	9S	147E
19	Lengai	Tanzania, Africa	8S	30E
20	Mauna Loa	Hawaii, USA	20N	155W
21	Mayon	Luzon, Philippines	12N	125E
22	Nevado Del Ruiz	Colombia	3N	76W
23	Ngauruhoe	North Island, New Zealand	38S	176E
24	Paricutin	Mexico	19N	108W
25	Pelee	Martinique	15N	61W
26	Pico Alto	Azores	27N	38W
27	Pinatubo	Philippines	18N	120E
28	Rabaul (several)	Papua New Guinea	4S	152E

EXPERIMENT 15.2 continued

PROCEDURE: continued

NUMBER	NAME	PLACE or REGION	LATITUDE	LONGITUDE
29	St. Helens	Washington State USA	46N	122W
30	Nea Kameni	Santorini, Greece	38N	23E
31	Soufriere	St. Lucia	13N	61W
32	Stromboli	Italy	38N	12E
33	Surtsey	Off Iceland	63N	20W
34	Taal	Luzon, Philippines	15N	121E
35	Tarawera	New Zealand	38S	176E
36	Mount Paektu	China	42N	130E
37	Vesuvius	Italy	41N	15E
38	Volcano	Italy (this is the original)	38N	14E
39	White Island	New Zealand	38S	177E
40	Yasur	Vanuatu	20S	170E

DATA and OBSERVATIONS:

Plot the location of each epicentre on the photocopy of the map (provided) using the numbers of each location. Students doing electronic reports will have to photograph or scan their completed map.

CONCLUSIONS:

1. Is there any pattern or logical groupings to the locations of the volcanoes?

2. Is there any relationship to the patterns seen with epicentres in Experiment 14.2?

3. Why do these volcanoes exist in these locations?

4. Account for the volcanoes in other locations where there might not be a grouping (e.g. Hawaii)?

5. How could a more accurate world pattern be obtained, especially in locations of isolated volcanoes?

Research (Optional)

Use the Internet to:

1. locate currently erupting volcanoes or those which have erupted in the last ten years.

2. locate any places where a major volcano MAY erupt e.g. a region with a large amount of heat which could cause the eruption of a new or existing dormant volcano.

Chapter 16: Geological Maps

EXPERIMENT 16.1 One or two lessons

MAPPING INCLINED BEDS

AIM: To draw a cross-section of inclined beds down to a suitable depth.

MATERIALS: Pencils, rulers, grid/graph paper, paper, protractors

BACKGROUND:

In Chapter One, students were shown how to make topographical cross-sections. Geologists can use a variety of techniques to find out what is happening to rocks and geological structures below the surface. The most basic and traditional way is by measuring the orientation of beds on the surface (dip and strike) and then using simple geometrical concepts to construct scaled drawings of how these beds may look like below the surface. Usually, the first part of these investigations involve surface mapping and drawing of topographical cross-sections. In the following exercises for simplicity, the surface will be considered to be flat.

PROCEDURE:

PART A. MAPPING DIPPING BEDS IN THE DIP DIRECTION where normally horizontal beds (symbol: +) have been tilted by Earth forces to a measureable angle of dip and the geologist has obtained data along the dip direction (at 90^0 to the strike which is North-South here).

Look at the PLAN VIEW (i.e. taken looking down onto the surface) of Map 1. on the next page.

1. Draw a topographical cross-section rectangle where the top is equal to the length A-B and represents the surface. Complete the rectangle so that it is about 6 cm deep.

2. Place the edge of a spare paper sheet along the map from A – B and mark off the locations of A, B and the boundaries of each bed. Also mark the angles of dip for each bed and the type of rock represented by the symbols.

3. Transfer this data to the drawn cross-section rectangle by placing the same paper edge between A and B and marking off the bed boundaries (including their dip).

4. Using a protractor along the top of the rectangle (= the land surface), draw in lines at the appropriate dip (say 40^0) in the dip direction (here, towards A) to represent he bedding planes.

5. Complete the cross-section by adding a small amount of rock symbol (as a representative of the whole bed) and the angle of dip for each bed.

EXPERIMENT 16.1 continued

PROCEDURE: continued

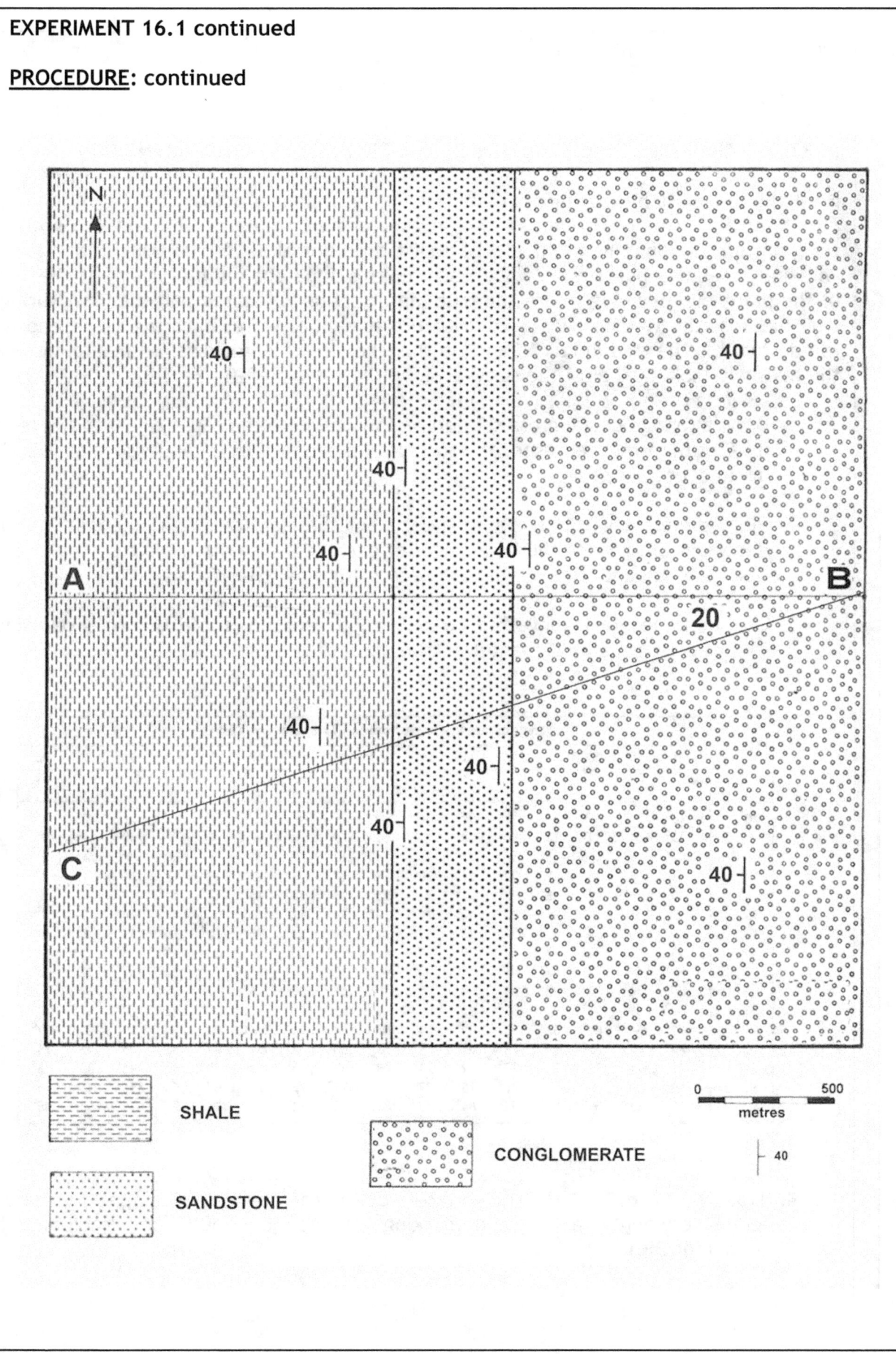

EXPERIMENT 16.1 continued

PROCEDURE: continued

PART B. MAPPING DIPPING BEDS NOT IN THE DIP DIRECTION (APPARENT DIP)

Here the geologist has taken data whilst walking obliquely across the dip direction in a direction of 20^0 (the direction angle) to it. Any angle of dip measurement will not be the true dip but an apparent dip if the geologist does not allow for it. Usually the geologist will find a bedding plane and compensate for this apparent dip by looking at the surface of the plane and deciding where the true dip direction is located and then measuring the true dip accordingly (one trick is to dribble a stream of water onto the plane and it will run downhill in the true dip direction). In this exercise, the geologist wishes to know what the beds will look like below the surface along the traverse B-C.

1. Draw a topographical cross-section rectangle where the top is equal to the length B-C as before to represent the surface. Complete the rectangle so that it is about 6 cm deep.

2. Place the edge of a spare paper sheet along the map from B - C and mark off the locations of A, B and the boundaries of each bed. Also mark the true angles of dip for each bed and the type of rock represented by the symbols.

3. Use the table which converts true dip to apparent dip (Table 16.1 of text) using the direction angle (20^0) and the true dip (40^0).

True Dip	Direction angle - angle between True Dip and Traverse															
	10°	15°	20°	25°	30°	35°	40°	45°	50°	55°	60°	65°	70°	75°	80°	85°
	Apparent dip															
10°	10°	10°	9°	9°	9°	8°	8°	7°	6°	6°	5°	4°	3°	3°	2°	1°
15°	15°	14°	14°	14°	13°	12°	12°	10°	10°	9°	8°	6°	5°	4°	3°	1°
20°	20°	19°	19°	18°	18°	17°	16°	14°	13°	12°	10°	9°	7°	5°	4°	2°
25°	25°	24°	24°	23°	22°	21°	20°	18°	17°	15°	13°	11°	9°	7°	5°	2°
30°	30°	29°	28°	28°	27°	25°	24°	22°	20°	18°	16°	14°	11°	9°	6°	3°
35°	35°	34°	33°	32°	31°	30°	28°	26°	24°	22°	19°	16°	13°	10°	7°	4°
40°	40°	39°	38°	37°	36°	35°	33°	31°	28°	26°	23°	20°	16°	12°	8°	4°
45°	45°	44°	43°	42°	41°	39°	37°	35°	33°	30°	27°	23°	19°	15°	10°	5°
50°	50°	49°	48°	47°	46°	44°	42°	40°	37°	34°	31°	27°	22°	17°	12°	6°
55°	55°	54°	53°	52°	51°	49°	48°	45°	43°	39°	36°	31°	26°	20°	14°	7°
60°	60°	59°	58°	58°	56°	55°	53°	51°	48°	45°	41°	36°	30°	24°	17°	9°
65°	65°	64°	64°	63°	62°	60°	59°	57°	54°	51°	46°	42°	36°	29°	20°	11°
70°	70°	69°	69°	69°	68°	67°	65°	63°	60°	58°	54°	49°	43°	35°	25°	13°
75°	75°	74°	74°	74°	73°	72°	71°	69°	67°	65°	62°	58°	52°	44°	33°	18°
80°	80°	80°	79°	79°	78°	78°	77°	76°	75°	73°	71°	67°	63°	56°	45°	26°
85°	85°	85°	85°	84°	84°	84°	83°	83°	82°	81°	80°	78°	76°	71°	63°	45°

4. Transfer this data to the drawn cross-section rectangle by placing the same paper edge between A and B and marking off the bed boundaries (including their Apparent Dip).

EXPERIMENT 16.1 continued

PROCEDURE: continued

5. Using a protractor along the top of the rectangle (= the land surface), draw in lines at the appropriate apparent dip in the new direction dip (here, towards C) to represent he bedding planes.

6. Complete the cross-section by adding a small amount of rock symbol (as a representative of the whole bed) and the angle of dip for each bed.

DATA and OBSERVATIONS:

Cross-sections for Parts A and B showing the dipping beds along these traverses.

CONCLUSIONS:

1. Why would a geologist NOT walk along the dip direction to measure true angle of dip?

2. What other geological or environmental situations cause difficulty in measuring the true dip of beds?

3. Why is dip direction a better parameter than strike when discussing the orientation of beds?

4. In Part A, at what depth would a drill core meet the top of the conglomerate if the drill was started on the surface along the traverse at the contact of the shale and sandstone (Hint: use the cross-section and its scale)?

Research (Optional)

Use the textbook or the Internet to revise how angle of dip and the dip direction is measured in the field.

EXPERIMENT 16.2 One lesson

MAPPING FAULTED BEDS

AIM: To draw a cross-section across a fault line and then measure the THROW and HEAVE of the fault.

MATERIALS: Pencils, rulers, grid/graph paper, paper, protractors

BACKGROUND:
There are several types of faults where the subsurface beds have been shifted up, down or horizontally along a fault line. Walking along a traverse, a geologist may suspect that there is a subsurface fault if there has been a sudden change in the beds such as one bed recurring further along the traverse. Subsurface mapping will show the parameters of the fault such as its throw (the vertical displacement) and the heave (the horizontal displacement) if they cannot be seen on the surface.

PROCEDURE:

Look at the PLAN VIEW (i.e. taken looking down onto the surface) on the next page.

1. Draw a topographical cross-section rectangle where the top is equal to the length A-B and represents the surface. Using an appropriate scale (say 1 cm = 100 metres depth), complete the rectangle so that it represents a depth of 600 metres.

2. Place the edge of a spare paper sheet along the map from A – B and mark off the locations of A, B and the boundaries of each bed. Also mark the angles of dip for each bed and the type of rock represented by the symbols and the position and dip of the fault line.

3. Transfer this data to the drawn cross-section rectangle by placing the same paper edge between A and B and marking off the bed boundaries (including their dip) and the position of the fault line.

4. Use a protractor to draw a line right through the cross-section at the dip (70°) of the fault line (as one does not know the depth of the fault and so assume that it goes beyond 600 m deep).

5. Using the protractor along the top of the rectangle (= the land surface), now draw in lines at the appropriate dip (say 40°) in the dip direction (here, towards A) to represent he bedding planes but DO NOT run the lines through the fault line previously drawn.

6. Now the base of any bed on the other side of the fault line must be estimated by finding its thickness (distance at 90° from top to bottom bedding planes for that bed) on the other side of the fault.

7. Complete the cross-section by adding any unknown bedding plane (use dotted lines) and then adding a small amount of rock symbol (as a representative of the whole bed) and the angle of dip for each bed.

EXPERIMENT 16.2

PROCEDURE: continued

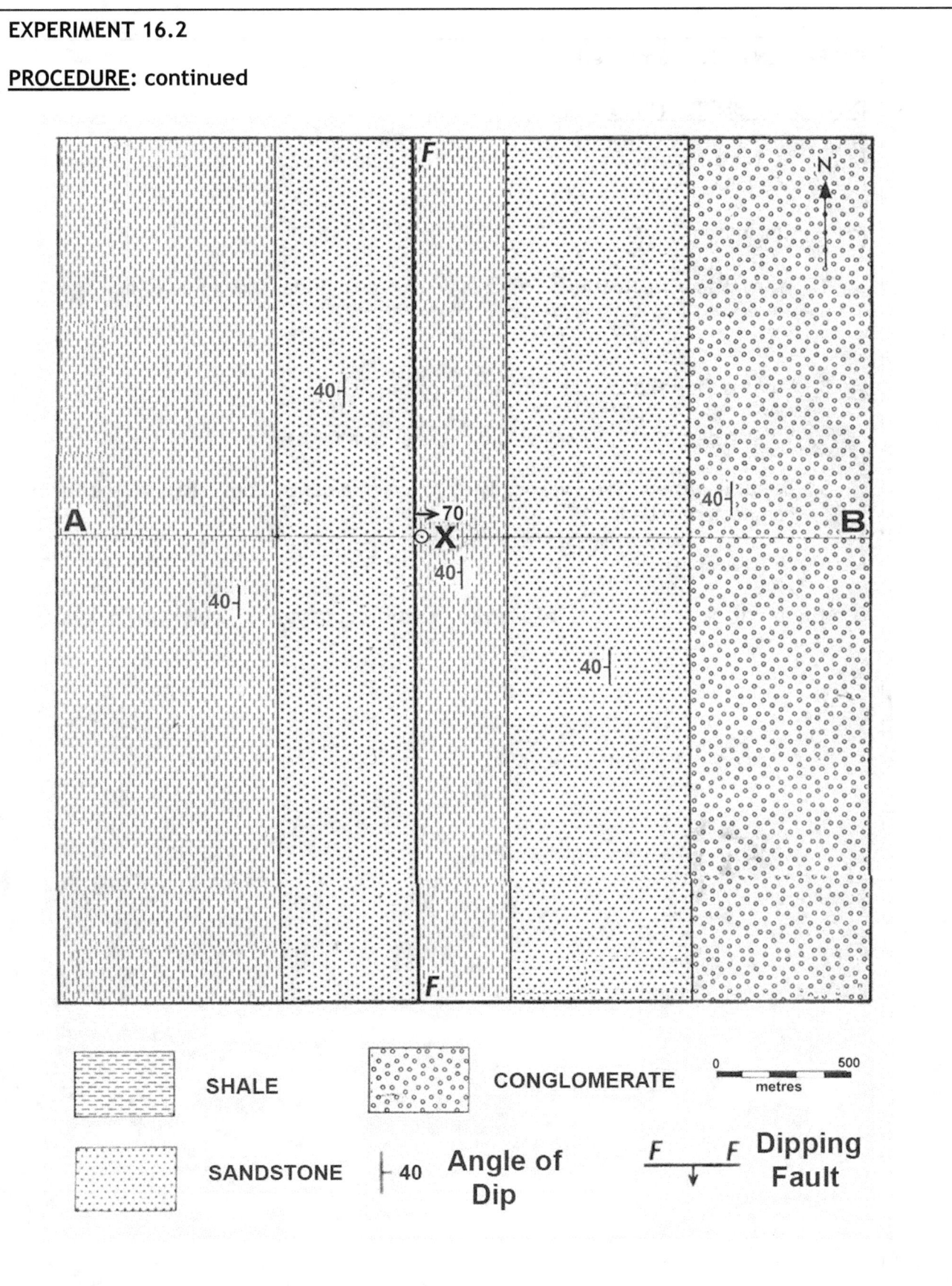

EXPERIMENT 16.2 continued

DATA and OBSERVATIONS:

Draw the cross-section for map showing the dipping beds along the traverse and the subsurface fault line.

CONCLUSIONS:

1. What type of fault is this? Which side of the fault has moved up or down?

2. Which side of the fault is the (a) hanging wall (b) the footwall?

3. What is the (a) heave and (b) throw of the fault?

4. What type of Earth forces would cause such a fault?

5. If a drill was put down vertically at X, at what depth would it strike the top of the conglomerate bed?

Research (Optional)

Use the textbook or the Internet to revise the different types of faults which undergo vertical movement.

EXPERIMENT 16.3   One lesson

MAPPING FOLDED BEDS

AIM: To draw a cross-section across folded beds.

MATERIALS: Pencils, rulers, grid/graph paper, paper, protractors

BACKGROUND:
Beds are usually folded by compressional forces. They can be simple upward folds (anticlines), downward folds (synclines) or a complex series of folds with equally angled limbs or sides (symmetrical) or with limbs of different angles of dip (asymmetrical).

PROCEDURE:

Look at the PLAN VIEW (i.e. taken looking down onto the surface) on the next page.

1. Draw a topographical cross-section rectangle where the top is equal to the length A-B and represents the surface. Using an appropriate scale (say 1 cm = 100 metres depth), complete the rectangle so that it represents a depth of 600 metres.

2. Place the edge of a spare paper sheet along the map from A – B and mark off the locations of A, B and the boundaries of each bed. Also mark the angles of dip for each bed and the type of rock represented by the symbols and the position of each fold axis (for simplicity, assume here that these are vertical).

3. Transfer this data to the drawn cross-section rectangle by placing the same paper edge between A and B and marking off the bed boundaries (including their dip) and the position of the fold axes.

4. Draw vertical dotted lines vertically right through the cross-section at each fold axis (as one does not know the depth of the fold and so assume that it goes beyond 600 m deep).

5. Using the protractor along the top of the rectangle (= the land surface), draw in lines at the appropriate dip in each of the different the dip directions to represent he bedding planes which have been folded and so dip in these directions. DO NOT run the lines through the fold axes lines previously drawn.

6. Now the base of any bed on the other side of the fault line may have to be estimated by finding its thickness (distance at 90^0 from top to bottom bedding planes for that bed) on the other side of a fold axis.

7. Complete the cross-section by adding any unknown bedding plane (use dotted lines) and then adding a small amount of rock symbol (as a representative of the whole bed but care with orientation of the symbol) and the angle of dip for each bed.

EXPERIMENT 16.3

PROCEDURE: continued

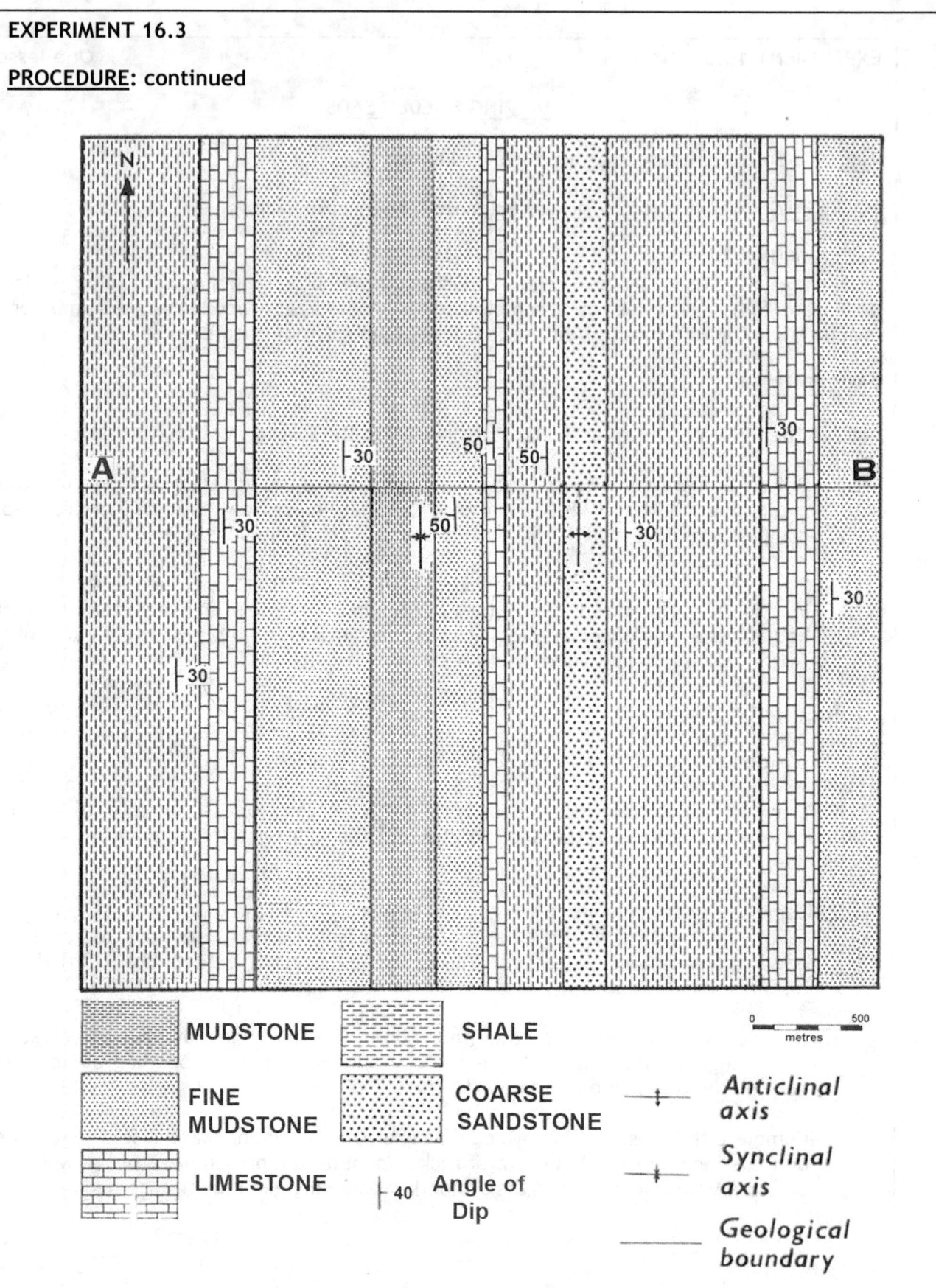

EXPERIMENT 16.3 continued

DATA and OBSERVATIONS:

Draw the cross-section for map showing the folding beds on each side of their axes.

CONCLUSIONS:

1. What type of fold is (a) closer to A and (b) closer to B?

2. What type of Earth forces would cause such folding?

3. Which sedimentary beds were deposited (a) first and (b) last?

4. When did the folding occur in relationship to the deposition of the beds?

Research (Optional)

Use the textbook or the Internet to revise the different types of folding.

EXPERIMENT 16.4　　　　　　　　　　　　　　　　　　　　　　　　One lesson

MAPPING IGNEOUS INTRUSIONS

AIM: To draw a cross-section across a region has been intruded by several igneous bodies.

MATERIALS: Pencils, rulers, grid/graph paper, paper, protractors

BACKGROUND:
As the name suggests, molten rock (magma) may come up through the exiting layers of rock (called the country rock) in several different forms. These intrusions may be small, thin tabular intrusions (dykes) which come up vertically along joints, cutting across existing rock layers or squeezed in between layers and are parallel to them (sills). They may also be extremely massive and come up vertically as rounded masses (as stocks) or even larger dome-shaped masses over many tens of kilometres across (as batholiths).

PROCEDURE:

Look at the PLAN VIEW (i.e. taken looking down onto the surface) on the next page.

1. Draw a topographical cross-section rectangle where the top is equal to the length A-B and represents the surface. Using an appropriate scale (say 1 cm = 100 metres depth), complete the rectangle so that it represents a depth of 600 metres.

2. Place the edge of a spare paper sheet along the map from A – B and mark off the locations of A, B and the boundaries of each bed. Also mark any angles of dip for each bed and the type of rock represented by the symbols and the position of each of the intrusions where they cut the surface. For simplicity, assume here that these are vertical with dykes being thin with parallel sides and stocks and batholiths coming up from a larger base.

3. Transfer this data to the drawn cross-section rectangle by placing the same paper edge between A and B and marking off the bed boundaries (including their dip) and the position of the edges of the intrusions.

4. Draw lines down through the cross-section for any dykes. If no dip is given for them assume that they are vertical, otherwise draw in the parallel sides of the dyke with the appropriate dip.

5. Draw in dotted lines (of uncertainty) down and spreading out from the surface edges of any stock or batholith to represent that they are a much larger mass below ground. Where these subsurface edges of these intrusions will not be known unless drill cores are taken.

6. Using the protractor along the top of the rectangle (= the land surface), draw in lines at the appropriate dip in each of the different the dip directions to represent any bedding planes which have been tilted and in their dip directions. Lines for tilted sedimentary beds can be carried through to the other sides of the intrusions.

7. Complete the cross-section by adding a small amount of rock symbol (as a representative of the whole bed but care with orientation of the symbol) and the angle of dip for each bed.

EXPERIMENT 16.4

PROCEDURE: continued

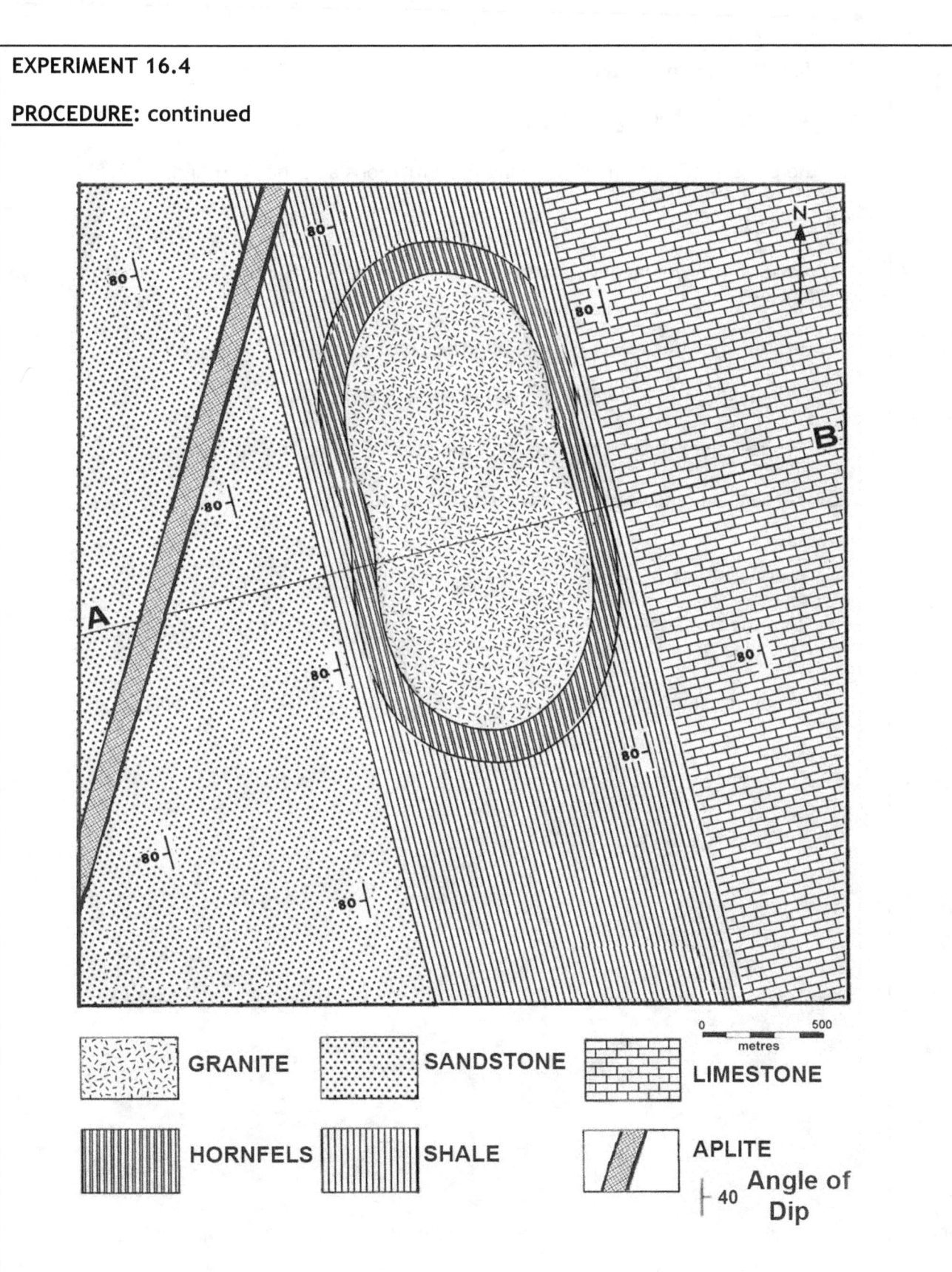

EXPERIMENT 16.4 continued

DATA and OBSERVATIONS:

Draw the cross-section for map showing the intrusions and the surrounding rocks.

CONCLUSIONS:

1. What types of igneous intrusions are seen on this map? Explain the reasons for your answer.

2. What is hornfels? Why does it ring the central body?

3. Why is there no ring or major edge along the dyke?

4. What is the relationship between the rocks granite and aplite?

5. Is there any indication which intrusion came first?

Research (Optional)

1. Use the textbook or the Internet to revise the different types of igneous intrusions.

2. What are some of the commercial uses for some igneous intrusions?

EXPERIMENT 16.5 One lesson

MAPPING UNCONFORMITIES

AIM: To draw a cross-section across a region which contains beds which are separated by a hiatus or break in time (called an unconformity)

MATERIALS: Pencils, rulers, grid/graph paper, paper, protractors

BACKGROUND:

The deposition of sedimentary beds is often not continuous. Sometimes there is a break in deposition caused by such events as lack of sediment or water flow followed by a period of erosion (often with some uplift of the basin floor). If a new cycle of sedimentation occurs in this area, an unconformity is formed.

PROCEDURE:

Look at the PLAN VIEW (i.e. taken looking down onto the surface) on the next page.

8. Draw a topographical cross-section rectangle where the top is equal to the length A-B and represents the surface. Using an appropriate scale (say 1 cm = 100 metres depth), complete the rectangle so that it represents a depth of 600 metres.

9. Place the edge of a spare paper sheet along the map from A – B and mark off the locations of A, B and the boundaries of each bed. Also mark any angles of dip for each bed and the type of rock represented by the symbols and the position of each of the intrusions where they cut the surface. For simplicity, assume here that these are vertical with dykes being thin with parallel sides and stocks and batholiths coming up from a larger base.

10. Transfer this data to the drawn cross-section rectangle by placing the same paper edge between A and B and marking off the bed boundaries (including their dip) and the position of the edges of the intrusions.

11. Draw lines down through the cross-section for any dykes. If no dip is given for them assume that they are vertical, otherwise draw in the parallel sides of the dyke with the appropriate dip.

12. Draw in dotted lines (of uncertainty) down and spreading out from the surface edges of any stock or batholith to represent that they are a much larger mass below ground. Where these subsurface edges of these intrusions will not be known unless drill cores are taken.

13. Using the protractor along the top of the rectangle (= the land surface), draw in lines at the appropriate dip in each of the different the dip directions to represent any bedding planes which have been tilted and in their dip directions. Lines for tilted sedimentary beds can be carried through to the other sides of the intrusions.

14. Complete the cross-section by adding a small amount of rock symbol (as a representative of the whole bed but care with orientation of the symbol) and the angle of dip for each bed.

EXPERIMENT 16.5

PROCEDURE: continued

EXPERIMENT 16.5 continued

DATA and OBSERVATIONS:

Draw the cross-section for map showing the unconformities by using a <u>wavy line</u> at the contact of the beds making the unconformity.

CONCLUSIONS:

1. What types of unconformities are shown in the cross-section?

2. What is different about the basalt? How did it get there?

3. Which beds were deposited (a) first and (b) last?

Research (Optional)

Use the textbook or the Internet to revise the different types of unconformities which can occur. Why are they all significant in determining the geological history of an area?

| EXPERIMENT 16.6 | One lesson |

GEOLOGICAL HISTORY

AIM: To draw a cross-section across a region and determine its geological history from the earliest events to the present.

MATERIALS: Pencils, rulers, grid/graph paper, paper, protractors

BACKGROUND:
Geologists use drawn cross-sections and surface geological maps (and much more) to determine what events have occurred in the region from the earliest to current. These will include the deposition of sedimentary beds, intrusions, metamorphic structures formed by Earth movements and erosion.

PROCEDURE:

Look at the PLAN VIEW (i.e. taken looking down onto the surface) on the next page.

1. Draw a topographical cross-section rectangle where the top is equal to the length A-B and represents the surface. Using an appropriate scale (say 1 cm = 100 metres depth), complete the rectangle so that it represents a depth of 600 metres.

2. Place the edge of a spare paper sheet along the map from A – B and mark off the locations of A, B and the boundaries of each bed. Also mark any angles of dip for each bed and the type of rock represented by the symbols and the position of any fold axes.

3. Transfer this data to the drawn cross-section rectangle by placing the same paper edge between A and B and marking off the bed boundaries (including their dip) and the position of fold axes.

4. Draw construction lines down through the cross-section for any fold axis.

5. Using the protractor along the top of the rectangle (= the land surface), draw in lines at the appropriate dip in each of the different the dip directions to represent any bedding planes which have been tilted and in their dip directions. Lines for tilted sedimentary beds can be carried through to the other sides of the fold axes.

6. Complete the cross-section by adding a small amount of rock symbol (as a representative of the whole bed but care with orientation of the symbol) and the angle of dip for each bed.

7. Review the entire completed cross-section and devise a geological history for all beds and structural events. Use all of the information learned about how specific rocks are formed and how geological events happen to include in this detailed history.

EXPERIMENT 16.6

PROCEDURE: continued

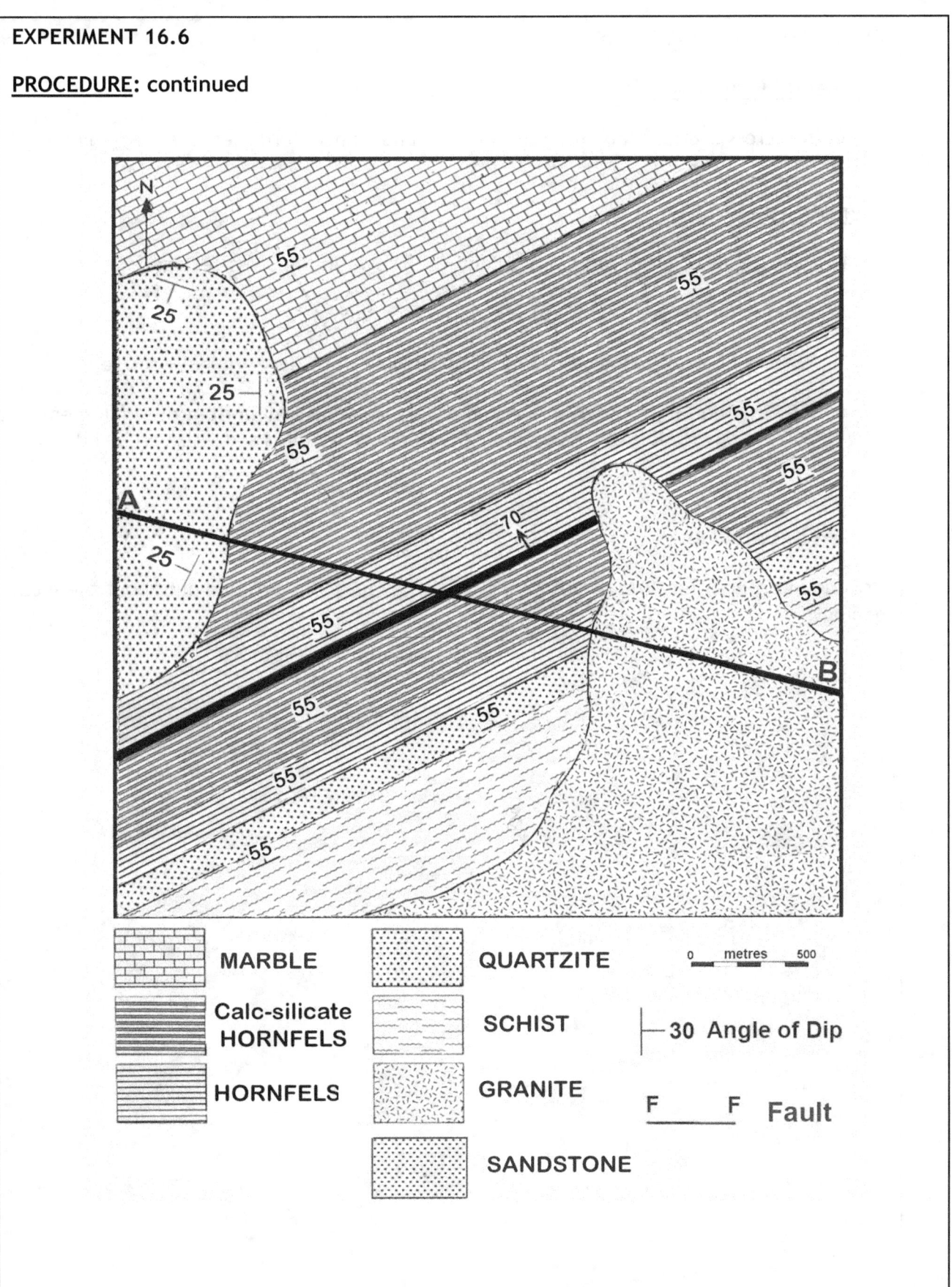

EXPERIMENT 16.5 continued

DATA and OBSERVATIONS:

Draw the cross-section for map showing, faults, unconformities (use a <u>wavy line</u>) and intrusions.

CONCLUSIONS:

1. What types of unconformity and fault are shown in the cross-section?

2. What is the (a) Heave and (b) Throw of the fault?

3. What is different about the layered beds? How were they originally formed?

4. LIST in order all of the events which occurred in this region from the first event to the current day (CARE: this is a complex problem so requires some research and in-depth description).

Research (Optional):

Use the Internet or local resources (maps, books, experts etc.) to find out about the local geology.

Chapter 17: Moving Sea and Wind

EXPERIMENT 17.1   One lesson

RELATIVE HUMIDITY

AIM: To construct a simple wet/dry bulb hygrometer and then measure the relative humidity.

MATERIALS: Thermometers (two/group), cotton gauze (or open weave cloth), small beaker or plastic container, stand with support, clock/stopwatch

BACKGROUND:

Relative humidity is the amount of water that the air can contain at that given temperature and is measured as a percentage. Evaporation rate depends on this amount with higher evaporation occurring at lower relative humidity. As water evaporates from material, it takes some of the heat from the material and drops its temperature. The amount that the temperature drops can be used to determine the relative humidity of the surrounding air.

PROCEDURE:

1. Take two thermometers and wrap a small amount of material gauze around the bulb of one of them.

2. Hang both thermometers from the support so that there is some separation between the two.

3. Wet the bulb with the gauze and hang the end of the gauze into the container now filled with water.

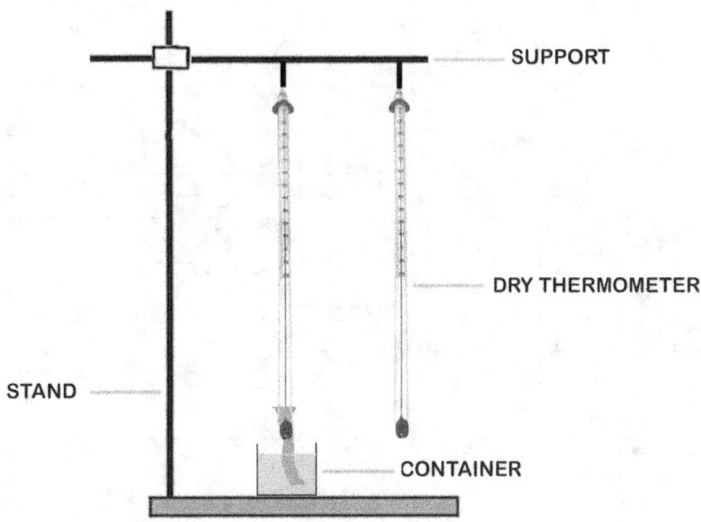

EXPERIMENT 17.1 continued

PROCEDURE: continued.

4. Stand the apparatus in a place with a free air circulation. If the classroom is air conditioned, continue with the experiment but then repeat it outdoors in the corridor or a shaded area.

5. Measure the temperatures of both thermometers every minute until there is no change in the thermometer with the wet bulb.

6. Taking the lowest temperature of the wet bulb and the difference between this and the temperature of the dry bulb, consult the following table to determine the relative humidity.

7. Leave the apparatus until near the end of the lesson (in the classroom) and re-calculate the relative humidity using the latest values shown on the thermometers. Is there a change?

DATA and OBSERVATIONS:

Set out a table showing TIME, DRY BULB TEMPERATRE and WET BULB TEMPERATURE.

Also show the temperatures and the relative humidity inside the classroom at the start of the experiment and near the end of the lesson. If appropriate show the temperatures and relative humidity outside and in the air conditioned classroom.

Wet-and-Dry Bulb Thermometer Relative Humidity

Dry Bulb Temp.	Dry Bulb Temperature minus Wet Bulb Temperature (zero difference =100% relative humidity)													
	1°C	2°C	3°C	4°C	5°C	6°C	7°C	8°C	9°C	10°C	12°C	14°C	16°C	18°C
10°C	88%	77%	66%	55%	44%	34%	24%	15%	6%					
11°C	89%	78%	67%	56%	46%	36%	27%	18%	9%					
12°C	89%	78%	68%	58%	48%	39%	29%	21%	12%					
13°C	89%	79%	69%	59%	50%	41%	32%	22%	15%	7%				
14°C	90%	79%	70%	60%	51%	42%	34%	26%	18%	10%				
15°C	90%	80%	71%	61%	53%	44%	36%	27%	20%	13%				
16°C	90%	81%	71%	63%	54%	46%	38%	30%	23%	15%				
17°C	90%	81%	72%	64%	55%	47%	40%	32%	25%	18%				
18°C	91%	82%	73%	65%	57%	49%	41%	34%	27%	20%	6%			
19°C	91%	82%	74%	65%	58%	50%	43%	36%	29%	22%	10%			
20°C	91%	83%	74%	66%	59%	51%	44%	37%	31%	24%	11%			
21°C	91%	83%	75%	67%	60%	53%	46%	39%	32%	26%	15%			
22°C	92%	83%	76%	68%	61%	54%	47%	40%	34%	28%	16%	5%		
23°C	92%	84%	76%	69%	62%	55%	48%	42%	36%	30%	19%	7%		
24°C	92%	84%	77%	69%	62%	56%	49%	43%	37%	31%	20%	9%		
25°C	92%	84%	77%	70%	63%	57%	50%	44%	39%	33%	22%	13%		
26°C	92%	85%	78%	71%	64%	58%	51%	46%	40%	34%	23%	14%	4%	
27°C	92%	85%	78%	71%	65%	58%	52%	47%	41%	36%	26%	16%	7%	
28°C	93%	85%	78%	72%	65%	59%	53%	48%	42%	37%	27%	17%	8%	
29°C	93%	86%	79%	72%	66%	60%	54%	49%	43%	38%	29%	20%	11%	
30°C	93%	86%	79%	73%	67%	61%	55%	50%	44%	39%	30%	20%	12%	4%
32°C	93%	86%	80%	74%	68%	62%	56%	51%	46%	41%	32%	23%	15%	8%
34°C	93%	87%	81%	75%	69%	63%	58%	53%	48%	43%	34%	26%	18%	11%
36°C	93%	87%	81%	75%	70%	64%	59%	54%	50%	45%	26%	28%	21%	14%
38°C	94%	88%	82%	76%	71%	65%	60%	56%	51%	47%	38%	31%	23%	17%

For example: The relative humidity with dry = 24°C and wet = 20°C is 69% at 24°C

EXPERIMENT 17.1 continued

CONCLUSIONS:

1) What is the relative humidity:

   (a) at the start of the lesson?
   (b) near the end of the lesson?

2) If there was any change, explain why this may have occurred?

3) If appropriate, was there any difference in humidity between an air conditioned room and outside? If so, why?

4) Explain why it is better to do the washing on a day with low relative humidity?

Research (Optional):

Use the Internet and class discussion to consider what household or industrial processes which require low relative humidity.

EXPERIMENT 17.2 One lesson

THE ANEROID BAROMETER

AIM: To construct a simple Aneroid Barometer and note any changes in air pressure.

MATERIALS: Large, open-mouth jars   rubber balloons or thin sheet rubber
            elastic bands   drinking straws   glue   stands   ruler

BACKGROUND:

Aneroid (without air) barometers work because the air outside changes whereas the small amount of air inside the box of the barometer does not. As the air changes outside, the relative air pressure between that inside the box of the barometer and that outside will cause a movement of the pointer of the barometer (see text).

PROCEDURE:

1) Obtain a large, open-mouthed jar which has a uniform rim.

2) Stretch some sheet rubber over the top (or use a rubber balloon) and secure this strongly with a rubber band and/or tape.

3) Glue the end of a drinking straw to the centre of the stretched skin and stand the apparatus so that the other end of the drinking straw is against a supported scale (e.g. a ruler)

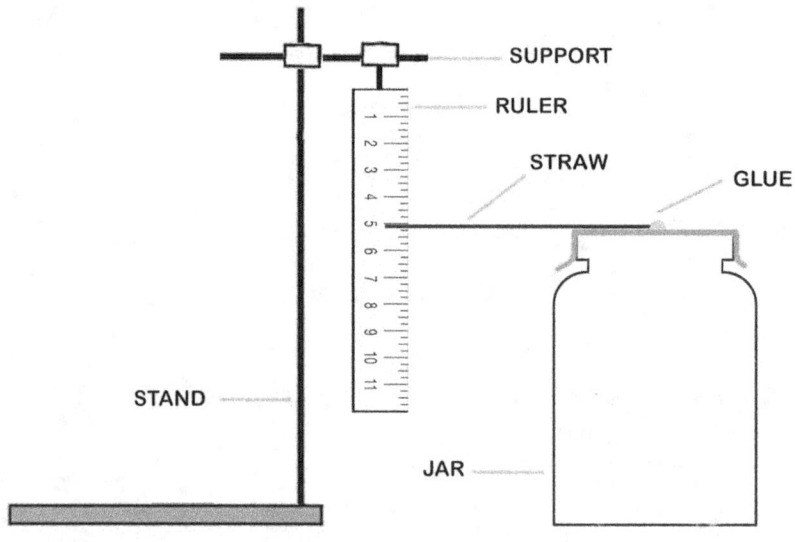

4) Measure any change in the position of the end of the drinking straw (and hence the relative air pressures between the air inside the jar and that outside) over several days.

EXPERIMENT 17.2 continued

DATA and OBSERVATIONS:

Describe what has happened over several days to the end position of the drinking straw.

CONCLUSIONS:

1. Did the relative air pressure change over several days? How?

2. What would be the errors in this experiment or the disadvantages of such a device for measuring air pressure?

3. How could this crude instrument be calibrated?

4. What would be the advantage of having a jar with less air pressure inside (as in the box of a real Aneroid Barometer)?

5. **Research (Optional):** Use the Internet to find out more about barometers, comparing the usefulness of Aneroid Barometers against Fortin Barometers.

EXPERIMENT 17.3   One lesson
WATER DENSITY

AIM: To describe what happens when waters of different densities due to heat or salinity meet.

MATERIALS: Class sets of small pill containers which have an eyedropper passed through their tops, beakers of icy cold water coloured blue, very hot water coloured red and saltwater coloured green.

BACKGROUND:

Within the ocean there are often many separate layers of water due to differences in density caused by differences in heat content and salt content (salinity).

PROCEDURE:

1. Obtain or make a pill container which has had an eyedropper pushed through its plastic top.

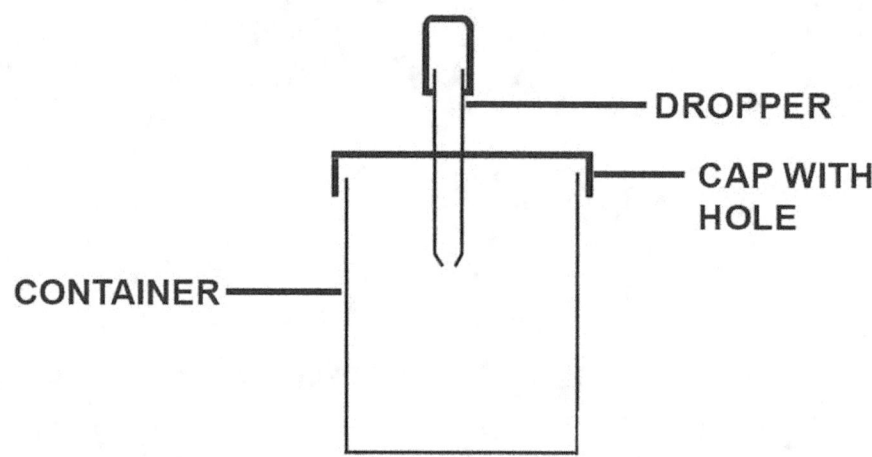

2. Fill the eyedropper completely with green, saturated salt solution. This is to be done carefully by removing the cap (including the dropper) and then squeezing the bulb of the dropper before placing it into the green-dyed saltwater. This represents seawater.

3. Completely fill the pill container with tap water. This represents freshwater (or seawater of a lower salinity).

4. Carefully place the cap and its dropper tightly onto the pill container being careful not to touch the bulb of the dropper. If any green liquid from the dropper goes into the container, wash it out and repeat the last three steps.

5. Turn the entire apparatus <u>sideways</u> so that the container and the nozzle of the dropper are horizontal.

EXPERIMENT 17.3 continued

PROCEDURE: continued.

6. Whilst observing the nozzle carefully, slowly squeeze the bulb of the dropper to push the green seawater into the clear freshwater.

7. Record the results and make a sketch of the apparatus in this horizontal position and the location of most of the green-dyed saltwater.

8. Open the container and completely wash it out. Use the container full of tap water to wash out the eyedropper by squeezing it out several times in the water.

9. Now fill the eyedropper with the hot water which has been dyed red. This represents a warm part of the ocean.

10. Repeat steps 2 to 6 (above) and record the results. A sketch may be drawn as in step 7 if required.

11. Completely wash out the container and eyedropper.

12. Now fill the eyedropper with icy water dyed blue. This represents a cold part of the ocean.

13. Repeat steps 2 to 6 (above) and record the results. A sketch may be drawn as in step 7 if required.

14. Continue to hold the container horizontal after squeezing all of the cold (blue) from the dropper. Wait for a while and see how the two waters mix.

15. Wash out and return all of the equipment including the beakers of dyed water.

DATA and OBSERVATIONS:

Describe what has happened in each of the three activities and draw appropriate sketches.

CONCLUSIONS:

1) What does this experiment suggest about the interaction between:

    (a) fresh and saltwater (or saltwater of different salinities)?

    (b) water of different temperatures?

2) Does the injected (cold) water instantly mix or does it stay separated for a while?

3) What parts of the ocean would be (a) warmer and (b) colder than the average ocean temperatures?

4) What processes (other than underwater volcanic action) would heat up seawater?

Research (Optional):

Use the Internet to find out more about Thermohaline Circulation?

EXPERIMENT 17.4 One lesson

CONVECTION CURRENTS

AIM: To observe a model of a convection current in water.

MATERIALS: A large (1000 ml) Pyrex beaker, tripod and Bunsen burner (no gauze mat), heat mat for desk, crystals of potassium permanganate (Condy's Crystals), tweezers.

BACKGROUND:

Within the oceans, waters of different temperatures move around the surface and also below by convection currents and other factors.

PROCEDURE:

CARE should be taken with the Bunsen burner, hot tripod and any hot water in the beaker when cleaning up. Allow the tripod and beaker to cool where they stand before cleaning and returning equipment.

1) Fill the large beaker almost to the top with cold tap water and stand it carefully on the tripod.

2) Light a Bunsen burner, turn it to the blue flame reduced to a medium size.

3) Carefully take one single crystal (if they are small there may be several in about a cubic millimetre) of permanganate and carefully drop it down the side of the beaker and allow it to fall to the bottom.

4) Move the flame of the burner until it is directly below the crystal in the beaker.

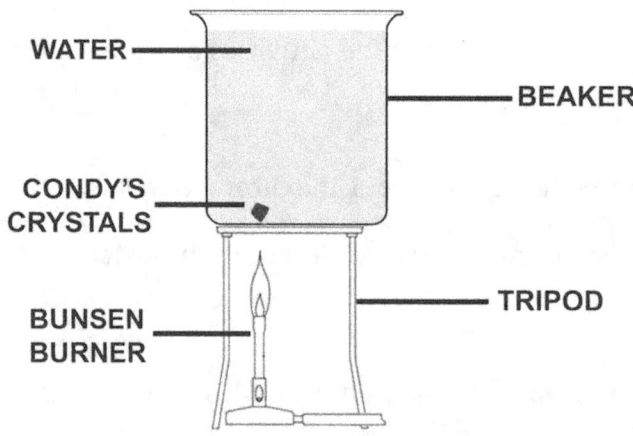

5) Observe carefully what happens to the crystal. Record the results and draw a sketch (similar to that above) showing what has happened after a time.

6) After the experiment, turn of the burner and allow the equipment to cool before washing the beaker and returning it and the rest of the apparatus.

EXPERIMENT 17.4 continued

DATA and OBSERVATIONS:

Describe what has happened when the water was heated around the Condy's Crystals. Make a sketch of the resulting pattern after it has formed.

CONCLUSIONS:

1) What happens to the water:

    (a) immediately above the crystal

    (b) at the surface of the water and

    (c) on the opposite side of the beaker from the burner (cool side)?

2) What is a convection current?

3) How do convection currents operate in the world's oceans?

4) Given an example of one of the great convection currents in the ocean by naming:

    (a) a warm ocean current and

    (b) a cold ocean current.

Chapter 18: Planet Earth

| EXPERIMENT 18.1 | One lesson or school or home time |

THE ROTATION OF THE EARTH

<u>AIM:</u> To find out how fast the Earth rotates by experiment.

<u>MATERIALS:</u>   metre rule   very large protractor
      shadow stick (a metre rule or a two metre long stick – the longer the better)

<u>BACKGROUND:</u> This method and its geometry were well-known to the Ancient Greeks and the principle was used in the construction of sundials.

<u>PROCEDURE:</u> (WARNING: Never look directly at the Sun)

1. Place a shadow stick of <u>known height</u> in the centre of an open area exposed to direct sunlight ensuring that the stick is <u>vertical</u> and placed on a large sheet of paper or on a flat surface which can marked or on which a shadow can be measured. Mark or note the first position of the base of the stick and along the line of the first shadow (or leave the shadow stick in place).

2. Measure the length of the first shadow.

3. Return in a few minutes (say every 15 min.) and note the position of the end of the new shadow.

4. Measure the length of this shadow.

5. Measure the angle (θ_1 to θ_3 etc.) between the line subtended from one shadow to the (e.g. from position 1 and the line now formed from the stick by the new shadow at position 2) AS ACCURATELY AS POSSIBLE using a large protractor.

6. Repeat steps 4 and 5 for each new shadow.

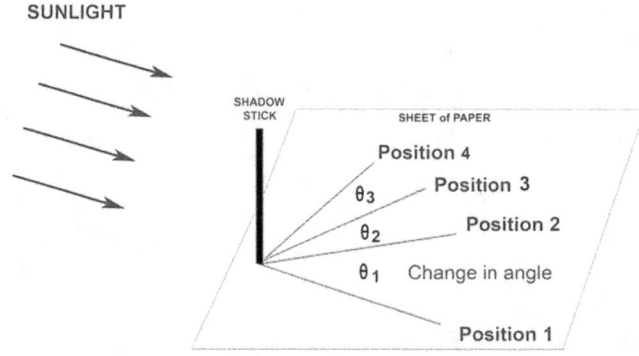

7. (GRAPH 1.) Graph the length of each shadow against the actual time (e.g. 10.00 am, 10.15, 10.30, 10.45, 11.00, 11.15, 11.30. 11.45, 12.00, 12.15 pm, 12.30, 12.45, 1.00 etc. depending upon time available in class or at home).

EXPERIMENT 18.1 continued

PROCEDURE: continued

8. (GRAPH 2.) Graph the CUMULATIVE angle (i.e. the continued addition of all angles from the start) against the actual times (as in step 7) using a time scale of 12 hours.

9. If measurements are extremely accurate, EXTRAPOLTATE the graph shape of GRAPH 2. To correspond to 12 hours and find the corresponding angle. This should give the rotation of the Earth for 12 hours so multiply by 2 and find how long does it take for one complete rotation (NOTE: this must be done VERY ACCURATELY and results often give large errors!)

Alternatively: One could also redraw the graph, extrapolating its shape to find the number of hours through which the shadow would pass as it turns through 360^0

DATA and OBSERVATIONS:

Record each of:

- Length of the shadow stick (centimetres):
- Length of the shadows in each position at each time:
- Each angle between each shadow:
- Time between measurements:

CONCLUSIONS:

1. Describe the shape of GRAPH 1.
2. What was the time for the shortest shadow (solar noon) from this graph?
3. Was this value exactly 12.00 noon? If not why was it different?
4. What was the shape of GRAPH 2? What does it say about the rotation of the Earth?
5. If the shape of GRAPH 2. was extrapolated, what was the angle or time for one rotation of the Earth. Calculate the percentage error for this calculation.
6. How could this measurement (step 9) be made more accurate?

Research (Optional)

Use the Internet to find out about

1. Sidereal time and how Earth time is measured today.
2. The instruments used by Tycho Brahe.

EXPERIMENT 18.2 One lesson

SIZE OF THE EARTH

AIM: To find the circumference of the Earth.

MATERIALS: metre rule   shadow stick (as before)   Internet and computer or tablet

BACKGROUND:

This method and its geometry were well-known to the Ancient Greeks, notably Eratosthenes of Cyrene (276-195 BC). Cyrene is now Aswan in southern Egypt and Eratosthenes was the Chief librarian at the great library at Alexandria in Egypt (then a Greek colony). He knew that at local noon on the summer solstice in Cyrene, the Sun was directly overhead because the shadow of someone looking down a deep well at that time blocked the reflection of the Sun on the water. He measured the Sun's angle of elevation at noon in Alexandria by using a vertical rod (a shadow stick or gnomon) to measure the length of its shadow. Knowing the length of the rod, and the length of the shadow, he calculated the angle of the sun's rays at Alexandria to be about 7° (1/50th of 360 ⁰ and so proportional to that part of the circumference of a circle). Assuming that the Earth as spherical (having seen its circular shadow on the Moon during a Lunar Eclipse), and knowing both the exact distance across the ground to Cyrene (a well-known trade route), he concluded that the Earth's circumference was fifty times that distance.

PROCEDURE: (WARNING: Never look directly at the Sun)

1. Place a shadow stick of known height in the centre of an open area exposed to direct sunlight ensuring that the stick is vertical.

2. Measure the length of the shadow along the ground.

3. From the length of the shadow stick and the length of the shadow, calculate the angle of the Sun by using trigonometry by:

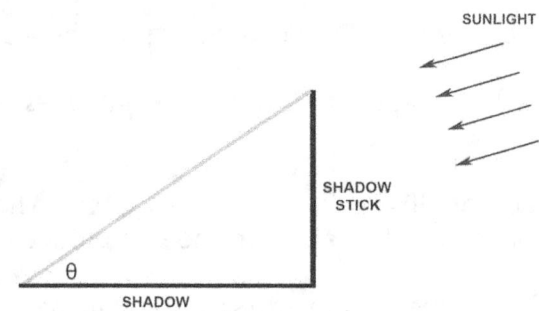

(a) finding the tangent of this angle (θ) by dividing the set height of the shadow stick by the length of the shadow.

(b) finding the inverse (tan^{-1}) of this value from tables or an Internet converter. (e.g. at https://www.rapidtables.com/calc/math/Tan_Calculator.html)

EXPERIMENT 18.2 continued

PROCEDURE: continued

4. Use an atlas or Internet maps to find another location North (or South in the Northern Hemisphere) which is on or near the Equator but with approximately the same longitude (i.e. is exactly due North).

5. Use a website to find this new locations Sun angle as approximately equal to the same time as the measured angle taken here.

    https://www.timeanddate.com/sun/australia/brisbane?month=3
    (type in the location of the place on the Equator)

6. Measure the distance (in kilometres) between the observers location and this new location on the Equator using a map or typing "distance between [x] and [y]" into a search engine where x and y are the two places of interest.

7. Find the difference in the angles of the Sun between these two positions. This corresponds to the angular difference between the two places at this known distance.

8. Use ratios to calculate the circumference of the Earth if this angular distance was 360^0

DATA and OBSERVATIONS:

Record each of the following:

> Height of the shadow stick (cm.)
> Length of the shadow (cm.)
> This location (latitude and longitude)
> Calculated Sun angle (degrees)
> Time of day
> Location of place on the Equator (give latitude and longitude)
> Angle of the Sun at this place on the Equator

CALCULATIONS:

Now the ratio of the angle difference ($\Delta\theta$) to 360^0 should be the same as the ratio of the land distance between the two places compared to the circumference of the Earth (C).

$$\text{i.e. } \Delta\theta / 360^0 = D/C$$

$$\text{or the circumference } C = 360 D / (\Delta\theta)$$

Where:

> ($\Delta\theta$) = Difference in Sun angles between the two places
> D = Distance between these two places (kilometres) and
> C = Circumference of the Earth (using North South radius)

The real value for the circumference around the Poles is 39,931 kilometres.

EXPERIMENT 18.2 continued

CONCLUSIONS:

1. What is the calculated value for the circumference of the Earth (along North South)?

2. How does this compare with the actual value of 39,931 kilometres? Calculate the percentage error for this experiment.

3. How could this experiment be improved?

Research (Optional): Use the Internet to find out:

1. how the actual size and shape of the Earth (Geodesy) has been determined with modern equipment.

2. more about Eratosthenes of Cyrene (Syrene)

Chapter 19: The Moon

EXPERIMENT 19.1   Several nights over one month

THE PATHWAY AND SHAPE OF THE MOON

AIM: To record the pathway of the Moon across the night sky and to record its shape over time.

MATERIALS:

- magnetic compass or other method of finding North (or South in the Northern Hemisphere)

- A good safe viewing area away from trees and tall buildings

- a paper diary or folder to record and keep drawings (ambitious students may also try photographs of the Moon in addition to drawings)

BACKGROUND:

The Moon, our only natural satellite has been the subject of recorded observations for thousands of years. This activity is an ongoing one requiring nightly (or at least regular) observations of the Moon's position and shape. Position might be difficult as the Moon sometimes rises and sets during the day. The main questions here are:

    (a) where does the Moon rise and
    (b) where does it set and
    (c) what is its approximate pathway across the sky?

The shape of the Moon should be easier to observe as there is some time between changes and so some nightly observations can be missed.

PROCEDURE:

1. At home, find a good, safe place from which the Moon can be observed. This place must be safe (NOT on a roof unless there is a wall or railing) and free of obstruction.

2. The direction of North (or south in the Northern Hemisphere) should be found using a compass, a star position (see the Textbook for finding North using Polaris or the South Celestial Pole) or a distinct landmark halfway between where the Sun rises and sets.

3. Allocate a set time each night (or every two or three nights) to observe the Moon.

4. Note its position in the sky with reference to some constant landmark (e.g. specific tree or building) and also the time of observation. Try to find out what time it rises and sets.

5. Note the relative size of the Moon as (a) it just rises (is high in the sky and (c) just sets.

EXPERIMENT 19.1 continued

PROCEDURE: continued

6. Also try to record the geographical bearing of the Moon's position. This can be done accurately with a magnetic compass or as a general reference as degrees left or right of the north point using hand angles
    (see: https://oneminuteastronomer.com/860/measuring-sky/)

7. Record the shape of the Moon, taking care to draw it accurately.

DATA and OBSERVATIONS: (may be a class effort)

Record each of the following:

- Time of rising and setting of the Moon at different dates
- General direction of these events (e.g. North, South, East, West)
- Maximum height of the Moon in the night sky (time, angle)
- Drawings of different shapes of the Moon and the date drawn.
- Dates of the first observation and last observation when the Moon returns to that same shape (a full Moon is best)

CONCLUSIONS:

1. Approximately where does the Moon rise and set?

2. Describe the Moon's general pathway across the sky. Why does it take this path?

3. What are the main changes in shape of the Moon (research text for appropriate names)? Give the order in which these shapes occur.

4. What causes the apparent change in shape of the Moon (use diagrams as required)?

5. Account for any apparent change in the SIZE of the Moon.

EXPERIMENT 19.2　　　　　　　　　　　　　　　　　　　　　　　　　　　**One lesson**

THE SIZE OF THE MOON

AIM: To use an ancient Greek method of estimating the size of the Moon.

MATERIALS:　　　　　Paper　Scissors　Drawing Compass
　　　　　Lunar Eclipse photographs (representing a direct view of the Moon)

BACKGROUND:

Aristarchus of Samos, one of the great astronomers of the Greek city in Egypt, Alexandria, used the time of passage of the Earth's shadow across the face of the Moon during a Lunar Eclipse to estimate the size of the Moon's diameter in 280 BC. This was calculated by estimating the number of Moon diameters the shadow of the Earth covered as it passed across the Moon's surface. This value was slightly larger than what is now known.

PROCEDURE:

1. From the printed copy of a photograph supplied on the next page (showing a multi-exposure of a lunar eclipse), measure the diameter of the full Moon on the photograph. <u>DO NOT MARK THE PAGE.</u>

2. Using a drawing compass, draw circles on a spare sheet of paper equal to the following multiples of the Moon's measured (scaled) diameter: 1½, 2, 2½, 3, 3½, 4 and 4½.

3. Cut out each of these shapes carefully on large sheets of paper.

4. Using at least two of the images of the Moon showing the Earth's shadow, find which of the cut circles will match the curvature of the arc of the Earth's shadow. (This will give the number of times that the Earth's diameter is greater than that of the Moon.)

EXPERIMENT 19.2 continued

PROCEDURE: continued

DATA and OBSERVATIONS:

Try each cut out shape to see which of the circles match up against the edges of the two partial Moons representing the shadow of the Earth on the Moon's surface during the eclipse.

Note down which of these multiples is the correct size (e.g. 1, ½, 2, 2 ½, 3, 3½, 4 and 4½ etc.)

CALCULATIONS:

Before an estimate was made (also by the Ancient Greeks) of the Earth's diameter, Aristarchus of Samos could only give the size of the Moon in terms of the fraction or ratio to that of Earth. Knowing (today) that the Earth's diameter averages at 12,756 km, estimate the diameter of the Moon by dividing the Earth's diameter by the determined fraction (e.g. 1, ½, 2, 2 ½, 3, 3½, 4 and 4½ etc.).

$$\text{Moon's diameter} = \frac{12{,}756 \text{ km}}{\text{scaled multiple in kilometres (i.e. 1, ½, 2 etc.)}}$$

EXPERIMENT 19.2 continued

CONCLUSIONS:

1. What is the calculated diameter of the Moon?

2. Compare the estimated value with the modern value (3,476 km).

3. Comment on the accuracy of this estimation.

4. What factors would give inaccuracies (consider astronomical, methodology and mathematical problems)?

Research: (Optional)

Use the Internet to find out about the life and contribution of Aristarchus of Samos

EXPERIMENT 19.3   **Night Observation**

THE MAIN FEATURES OF THE MOON

AIM: To observe the main features of the Moon.

MATERIALS: Sketch pad or photocopy of grid (below)
binoculars, small low-power telescope or similar

BACKGROUND:

The Moon has been observed for thousands of years and there had been much speculation as to the type of features found on the Moon - some of them inaccurate. Craters were seen and once thought to be of volcanic origin but now known to be due to meteor impacts. Mares, Latin for seas, were thought to be seas or oceans (perhaps dried) and are still named as such. These are now known to be flat lava plains.

PROCEDURE:

1. Choose a good night for viewing the FULL MOON over enough time to make all of the observations. Observe the Moon when it is relatively low on the horizon and a good size.

2. Use a gridded circle such as that below (use a photocopy – do not write on this book) to draw in the main features.

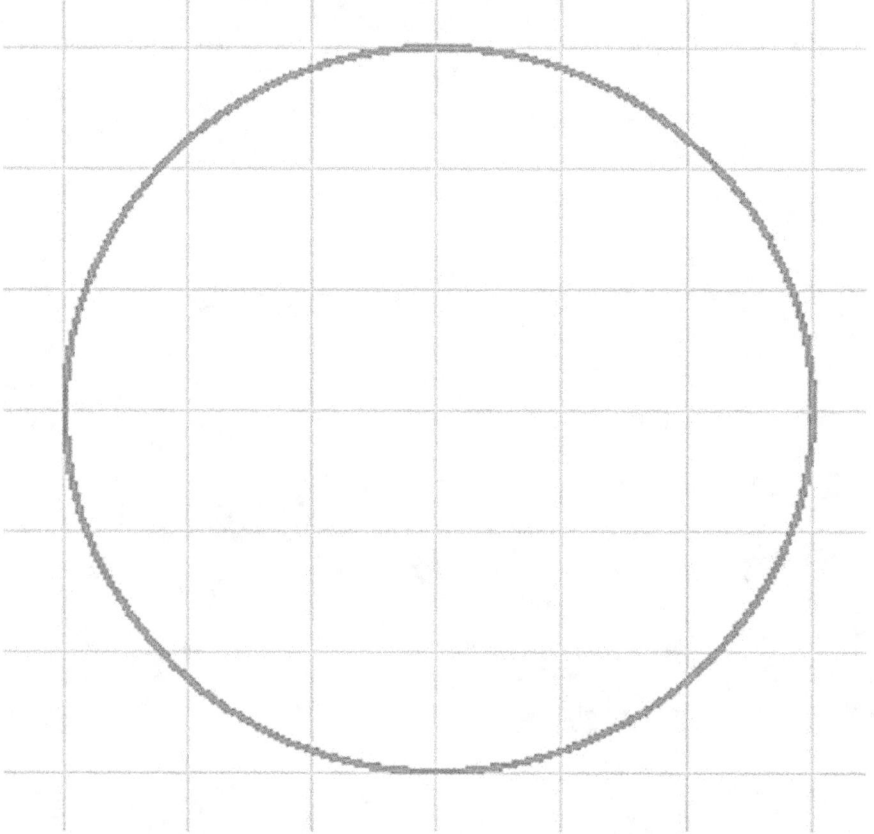

3. Use the textbook or the Internet to name the main features seen.

EXPERIMENT 19.3 continued

DATA and OBSERVATIONS:

Cut out and paste or scan and include the sketch made during the observation (including labels of the main features observed and their names).

CONCLUSIONS:

1. Why is the Moon larger at low angles?

2. What were the main features seen?

3. What aspect(s) of each feature enable its identification?

4. Comment of how the (a) craters and (b) the mares have been named.

Research: (Optional)

Use the Internet to find out about details of any of these features observed which have been more thoroughly described by actual observation ON the Moon by the Apollo Missions.

Chapter 20: A Matter of Perspective

EXPERIMENT 20.1 One lesson

PLANETARY MOTION

AIM: To determine the relationship between planetary velocities and distance from the Sun.

MATERIALS: Nylon or plastic twine   paper clips   stop watches   metre rules
smooth hollow tubes   rubber stoppers of equal weight

BACKGROUND: The planets orbit the Sun because the gravitational pull of the Sun and their velocities due to their original motion (produced at the beginning of the Solar System). Whilst the orbits of the planets are elliptical, they are near enough to a circle to allow this circular motion idea to be used. In this experiment, CENTRIPETAL FORCE (an inward force causing circular motion) applied by the hand represents the gravitational pull of the Sun and this is equated to the weight attached. A CENTRIFUGAL FORCE (an outward reaction force due to that force causing a circular motion) represents that of the planets due to their mass and velocity in their orbit.

In reality, the inward pull of the Sun is due to its mass and the distance from the Earth and Newton's Law of Universal Gravitation would apply such that this balances the outward force due to the planets mass, its distance from the Sun and its orbital velocity.

$$\text{i.e.} \quad GMm/r^2 = Mv^2/r$$

where G = Universal Gravitational Constant (6.67408×10^{-11} m^3 kg^{-1} s^{-2})
M = mass of the Sun
v = orbital velocity of the planet (m/s)
r = radius of orbit (distance to the Sun)
m = mass of the planet (kg)
g = acceleration due to gravity (9.8 m/s^2)

PROCEDURE: PART A: Orbital Period

1. Students are to work in pairs for this experiment.

2. With the twine passing through the tube, use the rule to measure out a radius of twine of 0.5 metres from the end of the rubber stopper on the end of the twine to be swung to the top of the tube.

3. Mark this distance by fastening the paper clip marker to the string JUST BELOW the bottom end of the tube.

4. Hold the tube in one hand and whirl the string and stopper around the head at a CONSTANT rate so that the paper clip stays in the same position just below the bottom of the tube (CARE: watch were the stopper is swinging. Look out for other students!).

EXPERIMENT 20.1 continued

PROCEDURE: continued

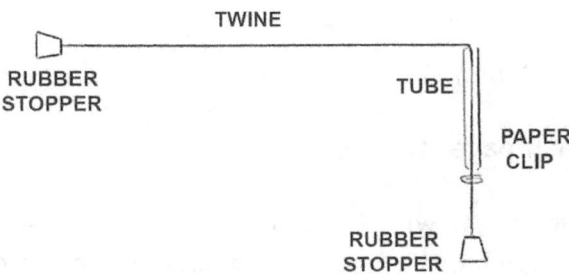

5. Whilst this is happening, the second student of the team measures the speed of the whirling stopper by counting the time (in seconds using the stop watch) for the stopper to make TEN revolutions and then dividing by ten to get the orbital period (time for one complete orbit in seconds.

6. REPEAT this experiment for radii of 1.5 m, 1.25 m, 1.0 m, 0.75m and 1.5 metres.

7. A graph of RADIUS OF ORBIT against ORBITAL PERIOD is drawn so that the relationship between these two factors can be seen.

Additional Activity: At the end of the measurements, have the mass swinging at a constant rate and the radius about 1.5 metres and then, holding the mass carriers, slowly pull them down.

PROCEDURE: PART B: Orbital Period and Velocities of the Visible Planets

1. Examine the following data for the six inner planets which are visible to the naked eye.

2. Graph their radius of orbit (distance from the Sun) against their orbital period.

3. Graph their radius of orbit against their orbital velocities

PLANET	ORBITAL RADIUS (Average - AU*)	ORBITAL PERIOD (years)	ORBITAL VELOCITY (km/sec)
Mercury	0.3871	0.2408	47.9
Venus	0.7233	0.6152	35.0
Earth	1.000	1	29.8
Mars	1.5273	1.8809	24.1
Jupiter	5.2028	11.862	13.1
Saturn	9.5388	29.458	9.6

* AU = the Astronomical Unit which is the average distance of the Earth to the Sun
= 149,597,871 km

EXPERIMENT 20.1 continued

DATA and OBSERVATIONS:

PART A:

Record measured data in a table, showing radius of orbit and times for orbital period.

Sketch the experiment (half page minimum).

Draw a graph showing own results with appropriate choice of axes, units and labels. Draw the line of best fit and determine the relationship between radius of orbit and orbital period.

Record any other meaningful observations

PART A:

Draw the two graphs separately with a suitable scale for each.

CONCLUSIONS:

1. From both graphs in Parts A and B for Radius of Orbit (r) against Orbital Period (T), what is the relationship between these two parameters?

2. What happens to the circular velocity of the rubber stopper when the mass carrier is slowly pulled down and the radius suddenly shortened? How does this relate to the orbits of the planets?

3. From the second graph in Part B, what is the relationship between the Orbital Radius and Orbital Velocities of these planets?

Research: (Optional)

1. Both the Geocentric and heliocentric Models of the Solar System appeared to work from an Earth-based observers' point of view. Compare and contrast both systems. Why did the heliocentric model become the accepted one?

2. Look up the equation for Newton's Law of Universal Gravitation and equate it to the centripetal force equation given in this experiment. Simplify this combined equation by removing parameters which are on both sides of the equal sign. What parameter will this new equation represent and why would this be useful to astronomers?

EXPERIMENT 20.2　　　　　　　　　　　　　　　　　　　　　　　One lesson

ELLIPTICAL ORBITS

AIM: To find the most accurate model of the shape of a planet's orbit around the Sun?

MATERIALS:　　drawing pins or ordinary pins　　large sheets of paper　　string　　pencil
　　　　　　　　　　　thick cardboard or foam　　　　ruler

BACKGROUND:

Johannes Kepler used the data that he and Tycho Brahe had obtained to determine the way by which the planets orbit the Sun. This was detailed in Kepler's three Laws of Planetary Motion which explained the motion of the planets in terms of elliptical orbits:

1. The orbit of a planet is an ellipse with the Sun at one of the two foci.
2. A line segment joining a planet and the Sun sweeps out equal areas during equal intervals of time.
3. The square of the orbital period of a planet is proportional to the cube of the semi-major axis of its orbit.

Later, Isaac Newton further extended the mathematics of these orbit shapes.

PROCEDURE:

1. Place a sheet of paper onto a thick piece of cardboard or a foam board and insert two drawing pins or ordinary pins in about the middle so that they are about 5 cm. apart (distance d).

2. Take a length of string (L) about 40 cm. and tie it into a loop and place this over both of the drawing pins.

3. 　　　With the drawing pins held firmly, place a pencil into the loop and pull the string outwards until it is tight.

4. With the string held tightly, move the vertical pencil around the pins to draw an ellipse.

5. Use the ruler to draw the major axis from the ellipse's sides through the foci of the drawing pins and the minor axis from the middle of the line through these pins and at right angles to it going out to the edge of the ellipse.

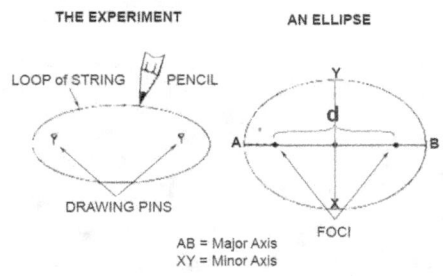

EXPERIMENT 20.2 continued

PROCEDURE: continued

6. Repeat steps 1 to 5 were repeated two or three times using different lengths of string (say 30, 20, 15) to get more ellipses with the same distance between the foci.

7. Repeat steps 1 to 5 were repeated two or three times using different distances between the drawing pins (say 4 and 3 cm) but the same length of string (say 15 cm).

8. Calculate the **eccentricity** (difference from a perfect circle) for each ellipse of step 6 (only) by measuring the distance from the centre of the ellipse to a focus (c) and to a vertex (a) from this focus to place on the ellipse's side where it meets the minor axis and dividing c by a:

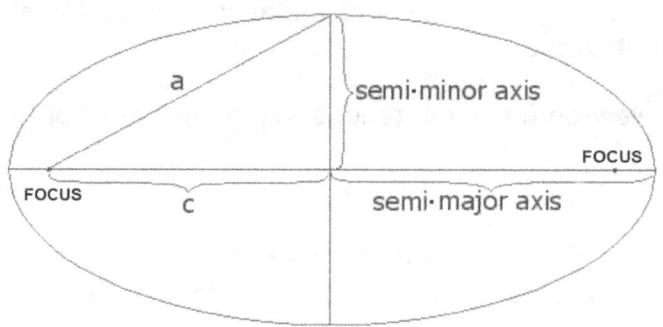

DATA and OBSERVATIONS:

Constructed Ellipse	Distance between pins (d) cm.	Length of string (L) cm.	Eccentricity e (=c/a)
1.			
2.			
3.			
4.			
5.			
6.			

Describe the changes which occurred as each variable (d and L) was changed.

EXPERIMENT 20.2 continued

CALCULATIONS:

Calculate the Eccentricities (e) for each of the ellipses in step 6 using the formula:

$$e = c/a$$

CONCLUSIONS:

1. What effect does the:
   (a) change in string length have on the shape of the ellipse?
   (b) change in distance between the foci have on the shape of the ellipse.

2. Earth has an eccentricity of 0.017 and that of Venus is 0.007.
   Which of these two orbits has an orbit more like a circle? EXPLAIN in terms of the eccentricity calculation.

3. In terms of the Earth's orbit, where is the Sun located in this model?

Activity: (Optional)

Given that the dwarf planet Pluto is distant from the Sun at 29.6 and 49.3 Astronomical Units (AU) respectively, and its eccentricity (e) is 0.2488, construct its elliptical to scale such that 1 cm = 5 AU to see what such an elliptical orbit for such objects resembles.

Hint: the Sun is NOT at the centre of the ellipse but at one of the foci (think about the set up. It is not like the diagrams above because the Sun cannot be at two foci) and that one will need to find the centre of the ellipse and its semi-minor axis which can be found from the formula:

$$b = h\sqrt{1 - e^2}$$

Where b = Semi-minor axis
h = semi-major axis and
e = Eccentricity

Chapter 21: More on Telescopes

EXPERIMENT 21.1   One lesson

THE REFRACTING TELESCOPE

AIM: To construct a simple refracting telescope and to describe its features, commenting on its usefulness in astronomy.

MATERIALS: Large convex lens   hand lens   metre rule

BACKGROUND:

Aids to extend vision first appeared in the 1500,s and in 1606 Hans Lippershey of Middleburg (in Holland) held a lens in front of another and noted that distant objects could be seen clearly. In 1609, Galileo Galilei, a professor of Mathematics at the University of Pisa (Italy) used this idea to construct a simple telescope and turned it towards the night sky. Galileo's observations confirmed the ideas of Copernicus and Kepler and introduced a new tool into the field of Astronomy

PROCEDURE:

1. Find the FOCAL LENGTH of each of the two lenses by focusing a distant object onto a sheet of paper. This is done by placing the lens in front of the paper and moving it back and forth until a clear image is formed. Measure the distance between the centre of the lens and the paper screen. Record these two values.

2. Look at each image on the paper and describe each separately in results. Use terms such as UPRIGHT or INVERTED (upside down); LARGE or DIMINISHED (smaller); REAL (can be projected onto a screen) or VIRTUAL (only seen if one looks through the lens).

3. Also describe the images seen by looking through each lens and record these descriptions.

4. CONSTRUCTING THE TELESCOPE - put the hand lens to the eye (the other may be closed or kept open) and then the bigger lens immediately in front of the eyepiece. Slowly move the bigger lens directly out from the eye until a clear image has been formed. Ensure that the two lenses are directly in line and vertical to this line of sight.

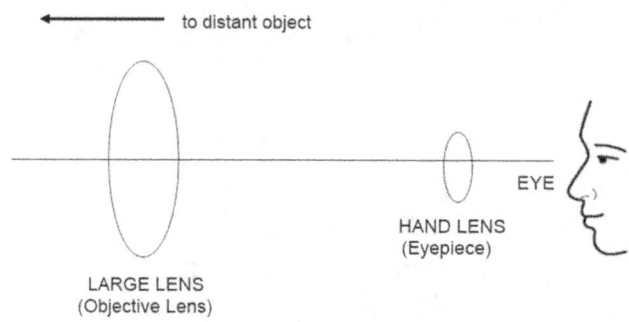

EXPERIMENT 21.1 continued

PROCEDURE: continued

5. Describe the nature (as in #2 above) of the image and record these descriptions.

6. ESTIMATE the MAGNIFICATION – the number of times that the image seems to be bigger (or smaller) than a vertical distant object from your position. To do this, locate a vertical object (chimney, lamp-post etc.) and focus on it. Keeping both eyes open, superimpose the image seen through the telescope over the original object seen by the other eye. Estimate how many times the image appears to be bigger than the object seen from your position. Record this value (e.g. as say X2 etc.).

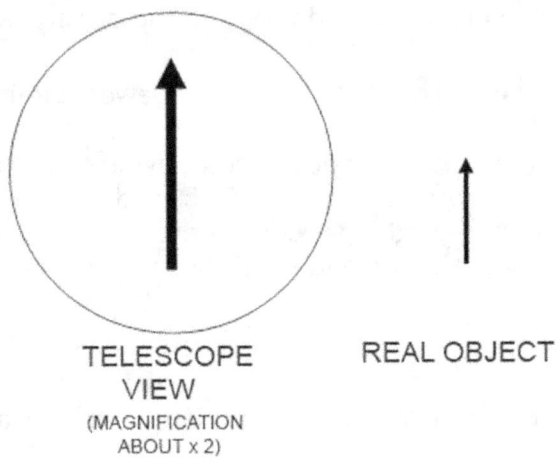

DATA and OBSERVATIONS:

Re-draw the following table and fill in the data:

	DIAMETER OF LENS	FOCAL LENGTH	NATURE OF REAL IMAGE	NATURE OF VIRTUAL IMAGE
OBJECTIVE LENS				
EYEPIECE				

Describe the nature of the image seen in telescope.

EXPERIMENT 21.1 continued

CALCULATIONS:

Magnification can also be calculated by dividing the focal length of the objective by the focal length of the eyepiece:-

$$\text{Magnification} = \frac{\text{FOCAL LENGTH OBJECTIVE}}{\text{FOCAL LENGTH EYEPIECE}}$$

Resolution is a measure of the clarity of image and the detail which can be seen. This is more important than magnification and is due to the quality and size of the objective lens:

Resolution (R) = 4.56/ D   (Dawes' Limit equation)

where D is the diameter of the objective lens in INCHES
(note: 2.54 cm = 1 inch) and
R is measured in seconds of arc (there are 60 seconds of arc in one minute of arc and 60 of these in one degree)

CONCLUSIONS:

1. How does the length of the telescope relate to the focal lengths of the two lenses in it?

2. Describe the orientation (vertical and horizontal position) of the image seen through this simple telescope.

2. Comment on any errors and disadvantages of such a telescope as the one constructed.

3. Describe the clarity of the image. How could it be improved?

4. What is (a) the magnification and (b) the resolution of the telescope ?

5. Comment on any problems/advantages in the construction of such a telescope.

6. Comment on the potential use of this telescope for astronomy.

Research: (Optional)

Use the Internet to:

1. Revise the discovery and use of refracting telescopes and the defects of lenses

2. Find out more about optical resolution.

EXPERIMENT 21.2   One lesson

THE REFLECTING TELESCOPE

AIM: To construct a simple reflecting telescope and to describe its features, commenting on its usefulness in astronomy.

MATERIALS: small concave mirror   hand lens   Metre rule

BACKGROUND:

Refracting telescopes were limited in size because of the difficulty in grinding large lenses. Remember that the bigger the Objective Lens, the better the resolution and the clear the image. In 1671, Sir Isaac Newton overcame the lens problem by using a silvered concave mirror as the objective to gather light. This was the first reflecting telescope (the Newtonian Reflector)

PROCEDURE:

1. Find the FOCAL LENGTHS of the mirror and hand lens by focusing a distant object onto a sheet of paper. For the mirror, this is done by facing the mirror towards the light and placing paper in front but to one side of the mirror. The paper is moved back and forth until a clear image is formed. Measure the distance between the centre of the mirror and the paper screen. The focal length of the lens is found by the same method as in the last experiment. Record these two values.

2. Look at each image on the paper and describe each separately. Use terms such as UPRIGHT or INVERTED (upside down); LARGE or DIMINISHED (smaller); REAL (can be projected onto a screen) or VIRTUAL (only seen if one looks through the lens).

3. Also describe the images seen by looking <u>through</u> the lens and into the mirror separately and record these descriptions in.

4. CONSTRUCTING THE TELESCOPE – face the mirror towards the distant light source. Put the hand lens to the eye (the other eye may be closed or kept it open) and then look into the mirror. Slowly move the lens directly out from the mirror until a clear image has been formed. Ensure that the mirror and lens are directly in line and vertical to this line of sight.

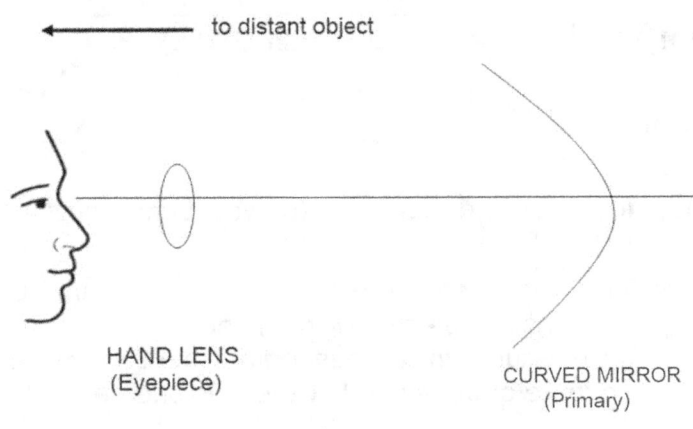

HAND LENS
(Eyepiece)

CURVED MIRROR
(Primary)

EXPERIMENT 21.2 continued

PROCEDURE: continued

5. Describe the nature (as in #2 above) of the image and record these descriptions.

6. Estimate the magnification – the number of times that the image seems to be bigger (or smaller) than a vertical distant object from your position. To do this, locate a vertical object (chimney, lamp-post etc.) and focus on it. Keeping both eyes open, superimpose the image seen through the telescope over the original object seen by the other eye. Estimate how many times the image appears to be bigger than the object seen from your position. Record this value (e.g. as say X2 etc.).

7. Calculate the theoretical magnification and resolution of the telescope:

DATA and OBSERVATIONS:

Re-draw the following table and complete the data recording:

	DIAMETER	FOCAL LENGTH	NATURE OF REAL IMAGE	NATURE OF VIRTUAL IMAGE
MIRROR				
EYEPIECE				

Describe the nature of the image seen through this telescope.

CALCULATIONS:

As in the previous experiment, magnification can also be calculated by dividing the focal length of the objective by the focal length of the eyepiece:-

$$\text{Magnification} = \frac{\text{FOCAL LENGTH OBJECTIVE}}{\text{FOCAL LENGTH EYEPIECE}}$$

Resolution is also found by:

$$\textbf{Resolution (R) = 4.56/ D} \quad \text{(Dawes' Limit equation)}$$

where D is the diameter of the Objective Mirror in INCHES
(note 2.54 cm = 1 inch) and
R is measured in seconds of arc (there are 60 seconds of arc in one minute of arc and 60 of these in one degree)

EXPERIMENT 21.2 continued

CONCLUSIONS:

1. How does the length of the telescope relate to the focal lengths of the mirror and the lens in it?

2. Describe the orientation (vertical and horizontal position) of the image seen through this simple telescope.

3. Comment on any errors and disadvantages of such a telescope as the one constructed.

4. Describe the clarity of the image. How could it be improved?

5. What is (a) the magnification and (b) the resolution of the telescope?

6. Comment on any problems/advantages in the construction of such a telescope.

7. Comment on the potential use of this telescope for astronomy.

Research: (Optional)

Use the Internet to find the advantages of using a reflecting rather than a refracting telescope in astronomy.

EXPERIMENT 21.3

SPECTROSCOPY

AIM: To use spectroscopes to examine the spectra of different elements.

MATERIALS:

- coloured pencils
- small, hand-held spectroscopes
- spectral tubes for some elements (Hydrogen, Helium, Sodium etc.)
- chemical salts of some common elements e.g. Strontium, Copper, Potassium, Sodium, Calcium and Bunsen burner, heat mat, paper clips and small beaker.
- Wiping towel

BACKGROUND:

A wide range of instruments can be attached to telescopes to analyse the light and other electromagnetic radiation coming from the Sun, stars and planets. In the late 17th Century, **Isaac Newton** discovered that a **spectrum** of colours can be made by passing sunlight through a triangular glass prism. Later, in the late 19th Century, scientists discovered that certain atoms, when heated or placed in **high voltage** electric currents, also gave characteristic **spectra** of colours. A **Spectroscope** is a device which uses a prism or diffraction grating to break up light into colours depending upon the elements emitting the light. It is possible to use a spectroscope to observe the coloured patterns due to individual elements in the laboratory but also on the Sun and other heavenly bodies.

PROCEDURE:

PART A: Electrical Discharge Tubes

1. Examine the spectroscope and note its features and construction.

2. CAREFULLY, look at the discharge tube spectra for EACH of the elements provided but do not get too close (CAUTION! HIGH VOLTAGE – DO NOT TOUCH SPECTRAL TUBES)

3. SKETCH the colours and relative positions of the spectral lines for each element provided.

PART B: Using powdered salts of some elements

1. Set up a heat mat and Bunsen burner giving the blue (invisible) flame.

2. Open out a paper clip to form a useful holder.

3. Dip the long end into water in a small beaker then into one of the powdered chemical salts. Note which one it is and record its name next to a strip into which the spectral lines will be drawn.

4. Carefully observe the spectrum of the element in this salt (CARE: Do not get too close to the flame and watch out for long hair!)

5. Carefully wash and wipe the end of the paper clip

EXPERIMENT 21.3 continued

PROCEDURE: continued

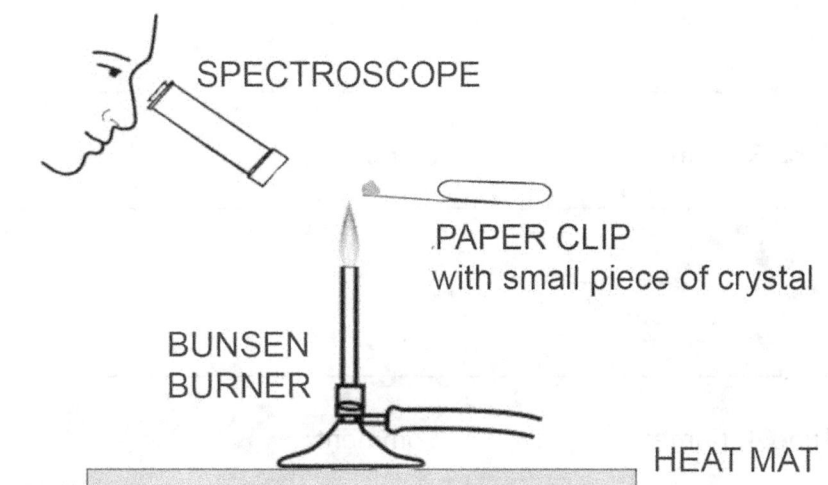

6. Repeat steps 3 to 5 for each of the other salts remembering to make a note of which element is being viewed and to wipe the paper clip clean after every use.

7. Turn off the Bunsen burner and clean the mat and desk. When the Bunsen is cold, also clean its base and shake any loose chemical out of its barrel (some can be unscrewed to be cleaned).

DATA and OBSERVATIONS:

Copy rectangular strips to represent the spectra of the discharge tubes, chemical element salts and the fluorescent lights and draw in the approximate positions of each spectral lines and their colours. DO NOT WRITE IN THIS BOOK.

PART A
Discharge Tube 1. Name: Comment:

EXPERIMENT 21.3 continued

DATA and OBSERVATIONS: continued

Discharge Tube 2. Name: Comment:

[ ]

Discharge Tube 3. Name: Comment:

[ ]

Discharge Tube 4. Name: Comment:

[ ]

Discharge Tube 5. Name: Comment:

[ ]

PART B:

Element 1. Name: Comment:

[ ]

EXPERIMENT 21.3 continued

DATA and OBSERVATIONS: continued

Element 2. Name: Comment:

Element 3. Name: Comment:

Element 4. Name: Comment:

Element 5. Name: Comment:

PART C: Fluorescent Light

Comment:

EXPERIMENT 21.3 continued

CONCLUSIONS:

Briefly describe the type or general appearance of the spectra seen in all of these activities

Research: (Optional)

Use the Internet to find the advantages of using a reflecting rather than a refracting telescope in astronomy.

List each of the main elements seen as spectra and give a brief description which could be used to describe its main characteristics (e.g. distinct colours and location of bands)

What element is in the fluorescent lights which give its characteristic spectrum (Hint: it is NOT Neon! Use the Internet to find a set of common spectra)

Explain how spectroscopy is a useful tool (attached to telescopes) in Astronomy.

Chapter 22: More on the Solar System

EXPERIMENT 22.1 One lesson

THE SOLAR SPECTRUM

<u>AIM:</u> To observe and describe the solar spectrum using a spectroscope.

<u>MATERIALS:</u>   Hand-held spectroscope

<u>BACKGROUND:</u>

The sun is a middle-sized, middle-age star which produces light, Heat and other radiation by nuclear fusion. It contains small percentages of the natural chemical elements but contains mostly Hydrogen plasma which is converted to helium. Newton first showed that the light from the Sun is composed of seven colours (those of the visible spectrum) but a spectroscope can break this light even further.

<u>PROCEDURE:</u>

1. Examine the spectroscope and note its features and construction.

2. Go outside and look near (BUT NOT AT) the Sun.

    NEVER LOOK DIRECTLY AT THE SUN

3. Observe in detail the spectrum produced by the SUN. Note any bright or dark lines seen in the spectrum and their approximate location.

4. SKETCH a representation of the solar spectrum.

<u>OBSERVATIONS AND DATA:</u>

Solar Spectrum (record observations in a box drawn like the one below. Do not draw in this book)

| |
| |

Also describe any detail seen.

<u>CONCLUSIONS:</u>

1. What type of spectra can be seen when looking at the Sun?

2. Some spectroscopes have numbers calibrated in them for the wavelengths of light. What are these for the solar spectrum (Internet or text search may be needed).

3. Explain the differences between a Continuous, Emission and Absorption Spectrum.

EXPERIMENT 22.2   **One lesson**

OBSERVING SUNSPOTS

AIM: To observe, sketch and possibly estimate the size of sunspots.

MATERIALS: Telescope with Sun Projector or alternative (given here)
pencil & ruler and plain paper

BACKGROUND:

Sunspots were reported by the Ancient Chinese about 800 BC and today can easily be seen using a telescope fitted with a solar projector which is simply a light guard with a hole through which the eyepiece tube fits and with a white screen mounted about 30 cm away. The alternative to a Sun Projector (is a simple telescope with a firm mount which can be set out in the yard. CARE - students should not attempt to view the Sun through the eyepiece! A large box (e.g. one which contained reams of photocopy paper or similar) should be firmly mounted so that the image from the eyepiece can be focussed onto a sheet of white paper place squarely at its base opposite the eyepiece.

PROCEDURE: (WARNING: Never look directly at the Sun)

1. Place a sheet of paper on the screen of the main telescope's solar projector or on the inside base of the small box which is placed over the eyepiece of the telescope (refracting or reflecting).

2. Aim the telescope towards the Sun (wear sunglasses if possible but still <u>do not look directly at</u>
3. <u>the Sun</u>). With the back to the Sun, align the telescope to the Sun by looking for the projected
4. image on the screen.

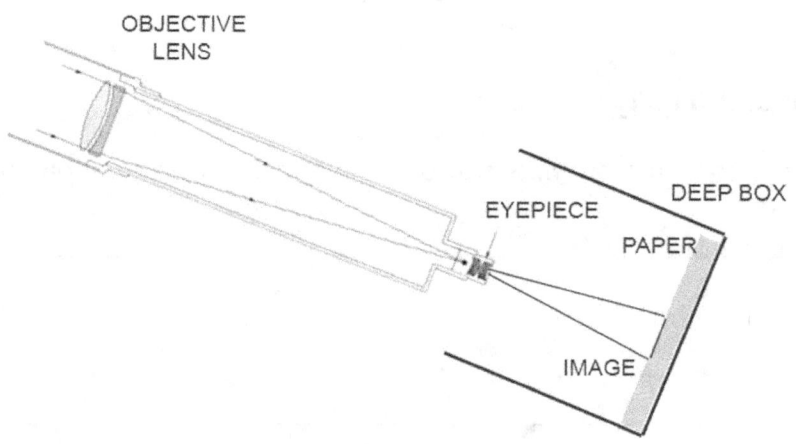

5. Each student or pair should then observe any sunspots on the screen, marking their position and size with a pencil. Also measure the diameter of the projected disk of the Sun (in mm.)

6. Returning to the classroom, students should then transfer their sketches onto a drawn circle representing the Sun's disk.

7. If any of the spots show some breadth, this should be measured with a millimetre scale.

EXPERIMENT 22.2 continued

OBSERVATIONS AND DATA:

Drawn numbers, positions and size of sunspots on circle (do not draw on this page)

Note the number of sunspots seen, the diameter (in mm) of the biggest one and their position on the Sun's disk.

Also note the diameter of the projected disk of the Sun's image

CALCULATIONS:

Use the ratio of the diameter of the disk projected (in mm.) to that of the widest sunspot and compare it to the diameter of the Sun (1.3914 million km) to find the real diameter of the sunspot i.e.

$$\text{Real Diameter (km.) of sunspot} = \frac{\text{measured sunspot (mm)} \times 1.3914 \times 10^6}{\text{measured diameter of projected image (mm)}}$$

CONCLUSIONS:

1. How many sunspots can be seen?

2. Are they all the same size?

3. What are the diameters of the biggest one on the drawing (in millimetres)?

4. Is there any preferred location for these spots on the Sun's surface (e.g. at the poles, at its equator etc.).

Research: (Optional)

Use the internet to find out about the true nature of sunspots and their activity cycle – referring to a time table of maximum sunspot activity.

EXPERIMENT 22.3　　　　　　　　　　　　　　　　　　　　**Several night observations**

INTRODUCTION TO THE PLANETS

AIM: To observe the planets in the night sky

MATERIALS:

- Good, safe viewing area of the night sky looking North (Southern Hemisphere) or South (Northern Hemisphere)
- Planet location App. on tablet or phone or data from the Internet as to what planets are up and where for the viewing location e.g.

    https://itunes.apple.com/us/app/star-planet-finder/id361753588?mt=8   for Apple

    https://play.google.com/store/apps/details?id=mobile.PlanetFinder.com for Android

    http://www.astroviewer.com/current-night-sky.php?lon=-73.94&lat=40.67&city=New+York+City&tz=EST   for PC tonight (set for closest location)

BACKGROUND:

For millennia, Humankind has observed that certain objects in the night sky did not stay fixed in groups (or constellations) like the stars but seem to wander. The Greek term *asters planetai* was used for these wandering stars and Roman names were later used for the planets which could be viewed with the naked eye. These were Mercury, Venus, Mars, Jupiter and Saturn. Not all will be up on any given night so viewing over time is required. Planetary data from the newspapers or using a star-finder app. or website can assist in viewing on a good night when planets will be visible. As Mercury orbits closer to the Sun. it will only be seen on rare occasions (when it is in its greatest elongation) only a few degrees from the Sun at sunset.

PROCEDURE:

1. Find a good, safe place from which to view the night sky looking towards the Equator and without many obstructions such as buildings and trees. Do NOT climb up onto unsafe structures to do so and if possible carry out the observations in pairs or a group.

2. Get a general idea of the night sky and observe some of the fixed patterns of stars (constellations) and then look for the planets for that night in the direction and angle indicated on the star-finder or chart. Confirm the identity of the planet(s) by using some simple descriptors e.g.
    - Venus is very bright and at a low angle after sunset
    - Mars will be higher and is a red colour
    - Jupiter will be bright and a pale yellow
    - Saturn is not as bright as Jupiter and is also yellow

3. Use hand angles vertically to find the angle above the horizon of the location of any planet and make a brief sketch of the position of the planet(s) in relative to the surrounding patterns of stars.

EXPERIMENT 22.3 continued

PROCEDURE: continued

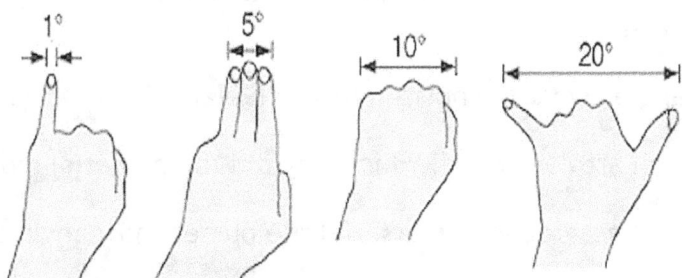

Rotate each through ninety degrees to measure vertical angles

4. Observe the location of the planet(s) over several days and note how it has moved in the sky compared to its background pattern of stars

OBSERVATIONS AND DATA:

Draw the approximate position of <u>each</u> planet and its star background on a semi-circle copied onto paper (do not draw on this page). Also mark East and West on this semi-circle (this may later be scanned for electronic recording of data). Several may be needed for viewing over several nights.

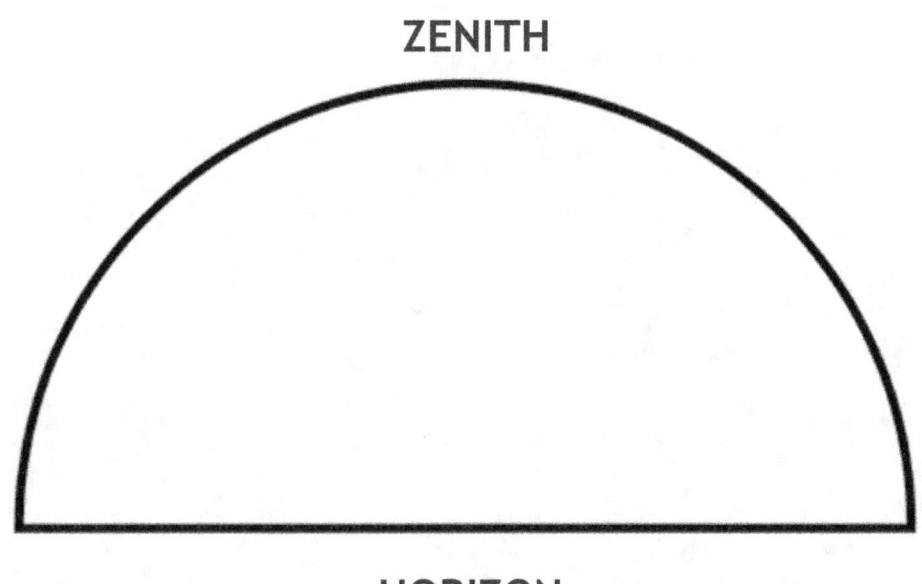

EXPERIMENT 22.3 continued

CONCLUSIONS:

1. What is different about the light (not the colour) coming from the planets compared to that of the stars?

2. What is the angle above the horizon for each planet?

3. Relative to the Earth's horizon, which way did the planet(s) move?

4. Relative to the background of stars, did the planet show any movement over one or two hours?

Research: (Optional)

Use the internet to find out about the life and work of Tycho Brahe and his assistant Johannes Kepler with special reference to the instruments which Tycho used.

EXPERIMENT 22.4 **One lesson**

THE ORBIT OF MARS

AIM: To construct a graphical representation of the orbit of the planet Mars using supplied astronomical data.

MATERIALS: Special graph paper planetary data

BACKGROUND:

Every two years or so, there is a short time over a couple of months when Mars' position from night to night seems to change direction and move backwards from west to east instead of its usual passage with the stars. This strange was a great problem for early astronomers trying to develop a satisfactory model for the orbit of Mars and the other planets around the Sun.

Positions of objects in the sky are often measured or plotted using the Celestial Coordinates (using the graduations on the mounts of telescopes) of Right Ascension and Declination. Right Ascension (R.A) is the the east-west coordinate of movement around the Celestial Equator (the projection of the Earth's Equator in the sky) as the angular distance east of the Vernal Equinox, measured in units of time (Hours and Minutes of Arc). Declination (Dec.) of an object is the is similar to latitude on the Earth and is measured in degrees from 0 to + 90 degrees north of the Celestial Equator, and 0 to -90 degrees south of it.

PROCEDURE:

On a large copy of the graph grid below, use the planetary data supplied to plot the position of the planet Mars over several months.

On each point plotted, mark the corresponding date on the graph.

OBSERVATIONS AND DATA:

For the years 1970-71

DATE	R.A.	Dec
31 Oct.	8 hr 46 min	19° 58 min
20 Nov.	9 19	18 12
20 Dec	9 47	16 56
9 Jan.	9 45	17 50
29 Jan.	9 23	20 30
18 Feb.	8 52	22 05
10 March	8 32	22 35
9 April	8 41	20 52
29 April	9 05	18 34
9 May	9 21	17 06

Data selected from: https://www.honolulu.hawaii.edu/instruct/natsci/science/brill/sci122/SciLab/L2/L2.html

EXPERIMENT 22.4 continued

OBSERVATIONS AND DATA: continued

Copy, do not draw on this graph.

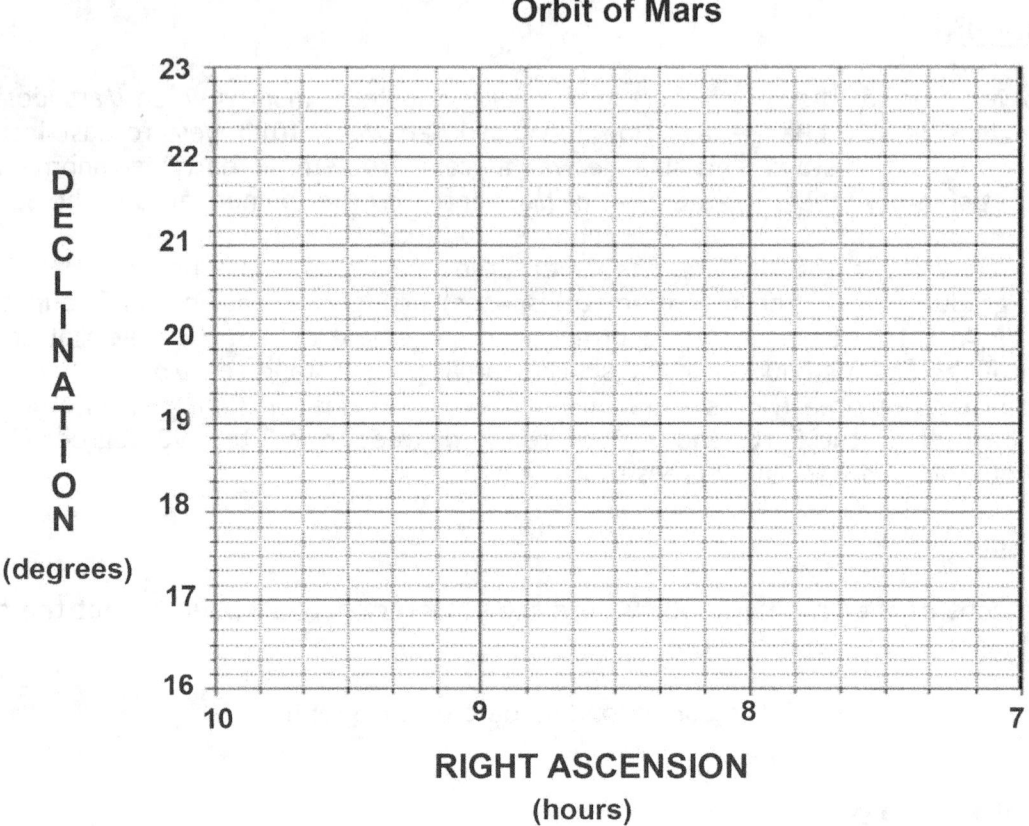

CONCLUSIONS:

1. Describe the shape of this orbit. What is this type of planetary motion called?

2. Over what time was this motion observed in the sky?

3. Describe what steps an astronomer would have to take to make such accurate measurements.

Research: (Optional)

Use the Internet to find out about:

1. Epicycles and who proposed this model for the orbit of Mars

2. How Johannes Kepler solved the riddle of this strange apparent motion using the Heliocentric Model of Copernicus?

Chapter 23: Beyond to the Stars

| EXPERIMENT 23.1 | Night Observation |

CONSTELLATIONS - SOUTHERN HEMISPHERE

AIM: To observe some of the main constellations of the Southern Hemisphere and describe some of their features.

MATERIALS: star chart or electronic star-finer app.   note book or iPad

BACKGROUND:

A **constellation** is a group of stars which form unchanging and meaningful patterns in the night sky. Ancient peoples have described them as typically representing animals, mythological people, gods, creatures, or in more modern times, manufactured devices. The International Astronomical Union recognises 88 modern constellations covering the entire celestial sphere. From the Earth, these constellations are seen as flat, two-dimensional patterns, but the stars within them can be vast distances apart and at different distances from Earth.

PROCEDURE:

1. Find a good, safe place from which to view the night sky looking towards the Equator (i.e. North) and without many obstructions such as buildings and trees. Do NOT climb up onto unsafe structures to do so and if possible carry out the observations in pairs or a group.

2. Get a general idea of the night sky and find some of the most common constellations which can be identified from a star finder. The best time for observation is in the Southern Hemisphere summer. Look for those constellations which have been described from the days of the ancient Greeks, remembering that they were described in the Northern Hemisphere, so here, they will appear upside down. Look for such constellations as Gemini (the twins), Canis Major (the big dog), Orion (the Hunter) and Taurus (the Bull). Near Taurus will be a star cluster called the Pleiades. In the winter, the most obvious constellations are the Crux (Southern Cross) and Scorpio.

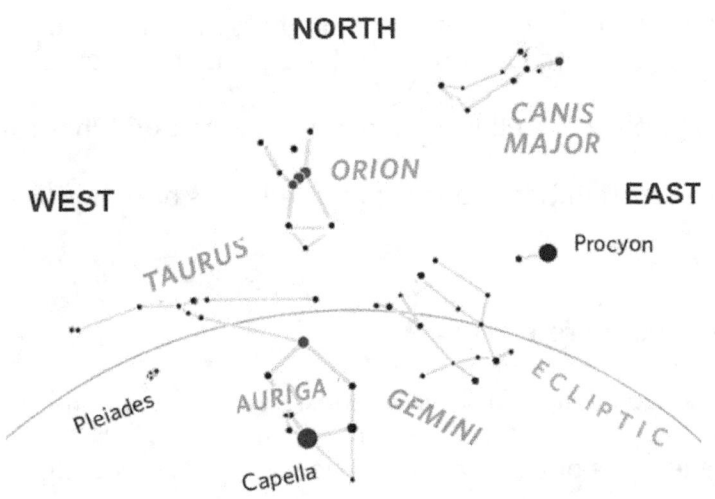

EXPERIMENT 23.1 continued

PROCEDURE: continued

3. Locate, describe and record in note form:

    a) The colour of the brightest star in Taurus and name it.
    b) The position of the red-orange star in Orion (sketch). Name it.
    c) The position of the brightest star in Canis Major (sketch). Name it.

4. Locate Crux (the Southern Cross) and on a sketch of it, label the brightness of each main star (as observed) in order of brightness from Alpha through Beta, Delta and Gamma, to Epsilon the dullest. Name them using these Greek letters.

5. Use the Southern Cross (low to the South in summer but higher in winter) to locate the South Celestial Pole and hence South on the horizon. Note any distinctive feature which is at South from the viewing position for later reference.

OBSERVATIONS AND DATA:

Copy sketches and data from the night observations, ensuring that objects are appropriately labelled. Include any photographs which may have been taken during the observation (NOT downloads).

Add any appropriate notes or descriptions.

CONCLUSIONS:

1. What is the colour and name of the brightest star in Taurus?

2. What is the position (e.g. bottom left etc.) and name of the red-orange star in Orion?

3. What is the position and name of the brightest star in Canis Major?

4. What is the position and name of the brightest star in Crux?
    i. (a sketch may be used for the last three questions)

5. Which way do the constellations move across the sky? What is this pathway called?

6. Why do the constellations appear to move in this direction?

Research: (Optional)

a) Use the Internet to research:

    a) the Greek legions of the Pleiades, Taurus, Orion and the Scorpion.

    b) the legends of other peoples about common constellations.

    c) How one can find North using Orion.

b) Use the constellation of Orion framework to design a new outline and story.

| EXPERIMENT 23.2 | Night Observation |

CONSTELLATIONS - NORTHERN HEMISPHERE

AIM: To observe some of the main constellations of the Northern Hemisphere and describe some of their features.

MATERIALS: star chart or electronic star-finer app.   note book or iPad

BACKGROUND:

A **constellation** is a group of stars which form unchanging and meaningful patterns in the night sky. Ancient peoples have described them as typically representing animals, mythological people, gods, creatures, or in more modern times, manufactured devices. The International Astronomical Union recognises 88 modern constellations covering the entire celestial sphere. From the Earth, these constellations are seen as flat, two-dimensional patterns, but the stars within them can be vast distances apart and at different distances from Earth.

PROCEDURE:

1. Find a good, safe place from which to view the night sky looking towards the Equator (i.e. South) and without many obstructions such as buildings and trees. Do NOT climb up onto unsafe structures to do so and if possible carry out the observations in pairs or a group.

2. Get a general idea of the night sky and find some of the most common constellations which can be identified from a star finder. The best time for observation is in the Northern Hemisphere summer.

3. Looking South and high in the sky, notice three bright stars to the East. These are Vega, Deneb and Altair which form the informal summer triangle. Notice the constellations which contain these three bright stars and sketch these constellations and give their ENGLISH names (i.e. what the constellations and their ancient names represent).

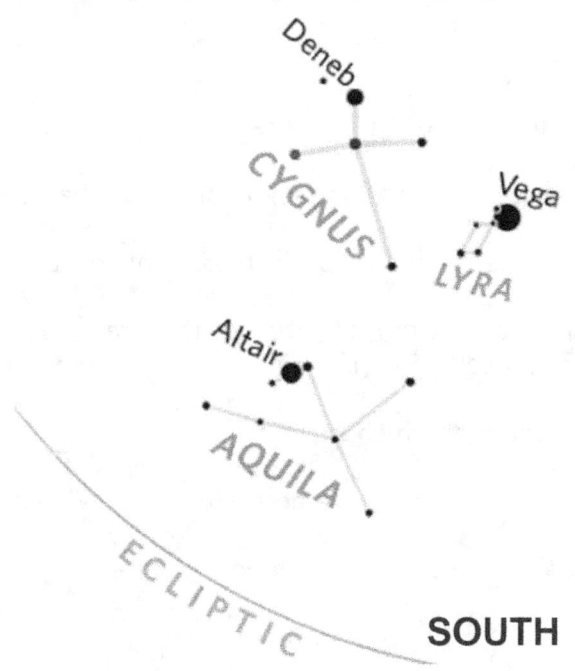

EXPERIMENT 23.2 continued

PROCEDURE: continued

4. Facing north and looking high in the sky, locate the Big Dipper (also called the Plow or the wagon in parts of Europe). Follow the long tail or handle of the Dipper and find Polaris, the North Star. Look quickly down to the horizon and find an object on it which is now north of the observation position. Remember this for future reference.

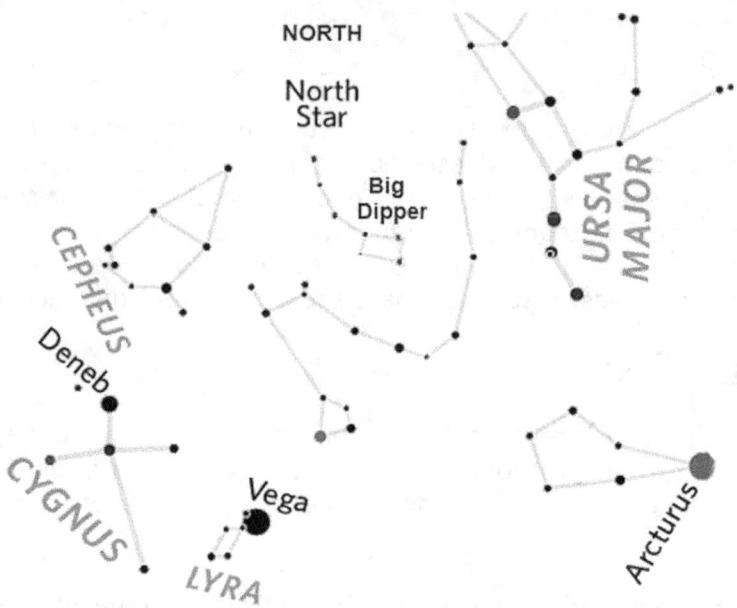

OBSERVATIONS AND DATA:

Copy sketches and data from the night observations, ensuring that objects are appropriately labelled. Include any photographs which may have been taken during the observation (NOT downloads).

Add any appropriate notes or descriptions.

CONCLUSIONS:

1. What is the colour and name of the brightest star in Cygnus?

2. What is the brightest star in the "Summer Triangle"? What constellation is it in?

3. Describe the shape of the constellation just north of this brightest star. What is its name?

4. What is the position (e.g. bottom left etc.) of the orange star in Aquila? What is its order of brightness in this constellation (i.e. alpha, beta, gamma etc.)?

5. What is the name of the star which is the brightest in the constellation of Ursa Minor?

6. Which way do the constellations move across the sky? What is this pathway called?

7. Why do the constellations appear to move in this direction?

EXPERIMENT 23.2 continued

Research: (Optional)

Use the Internet to:

1. Find out how the names Cygnus, Lyre, Aquila and Ursa Minor were derived?

2. What is meant by the term circumpolar star?

EXPERIMENT 23.3	One period

FINDING DISTANCE BY PARALLAX

<u>**AIM:**</u> To construct a rangefinder and then use it to measure a distance by the method of parallax.

<u>**MATERIALS:**</u>   Length of board (about 3 cm x 1 cm x 1500 cm)   tape measure
large cut-out protractor   drinking straws   pins small hammer or metal weight

<u>**BACKGROUND:**</u>

Parallax (from Ancient Greek *parallaxis* meaning alternation) is the difference in the apparent position of an object when viewed along two different lines of sight and is measured by the angle of inclination between those two lines. Nearby objects show a larger parallax than farther objects when observed from different positions, so parallax can be used to determine distances. To measure large distances, such as the distance of a planet or a star from Earth, astronomers use the principle of parallax with two sight-lines to the star observed when Earth is on opposite sides of the Sun in its orbit.

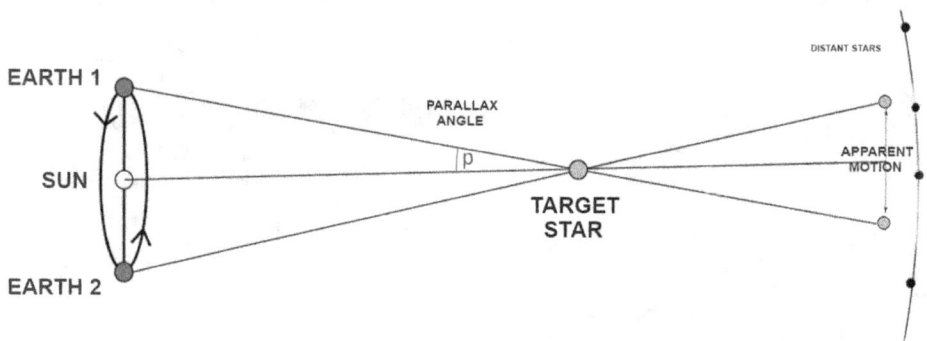

<u>**PROCEDURE:**</u>

1. Hold the finger upright and at arm's length and look at a distant object with only one eye then the other to see the parallax shift of the object against its background.

2. To construct a parallax angle rangefinder, first glue cutout photocopies of the large protractors onto the end of a thin piece of board about 1.5 to 2 metres long, ensuring that the true base of the protractor is on the edge of the wood.

3. Next, use pins to pierce through two drinking straws and then hammer them into the board at the zero part of the protractor (where its base meets the vertical line for $90°$). These drinking straws should be free to rotate (but not too freely) and should be flat on the protractor. These will be the sights through which the distant object will be viewed.

4. Finally, raise this rangefinder onto a small laboratory stool or box so that it will be convenient to look through the sights at a target object on the other side of the room.

EXPERIMENT 23.3 continued

PROCEDURE: continued

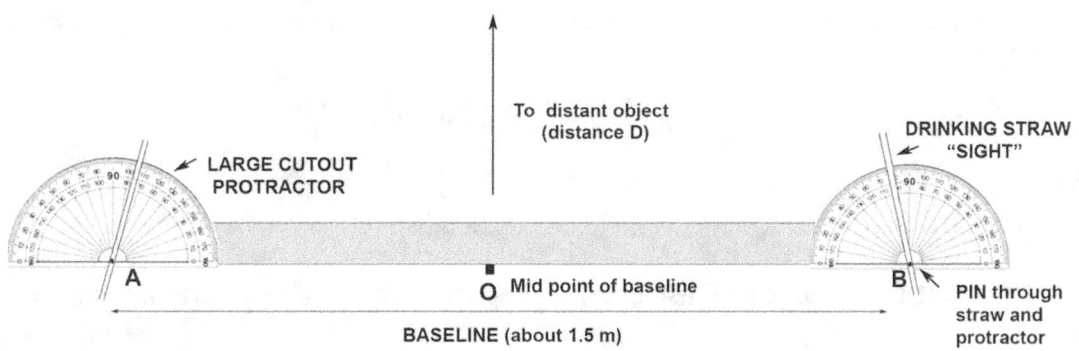

5. With the rangefinder facing the distant object (at 90^0 to the line of sight towards the object) and firmly held, line up one sight on the centre of the object and measure its angle on the protractor.

6. Repeat this with the other sight whilst keeping the rangefinder board perfectly still.

7. Calculate the distance to the object using the method of parallax and compare it to the measured distance to the object.

(NOTE: In this activity steps 1 and 5 to 7 can be attempted with more distant objects but the angles will be smaller and the error will be larger).

OBSERVATIONS AND DATA:

Measure and record the length of the baseline AOB (in metres). Record the each of the sighted angles (to the nearest 0.5 of a degree). Measure the real distance between the centre of the rangefinder (at O) and the distant object with a tape measure.

CALCULATIONS

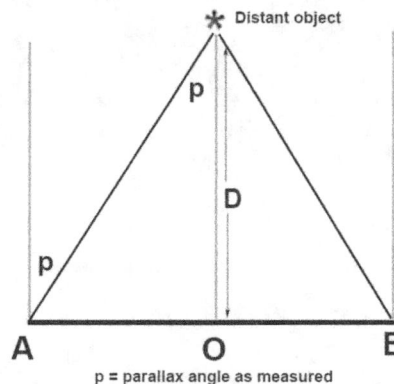

p = parallax angle as measured

Now the TANGENT of p will be   AO/ D

Where AO is one half of the baseline length and

So Distance to the object (D) = AO/Tan P

(tangent calculator is at
https://www.rapidtables.com/test/testt.html)

EXPERIMENT 23.3 continued

CONCLUSIONS:

1. What was the calculated distance using the method of parallax?

2. How did this compare with the real (measured) distance? What was the percentage error?

3. How could this experimental method be improved?

Research: (optional)

Use the Internet to find out how astronomers measure the angle of parallax.

www.ingramcontent.com/pod-product-compliance
Lightning Source LLC
Chambersburg PA
CBHW080128110526
44587CB00019BA/3364